Technology, health, and the patient consumer in the twentieth century

Manchester University Press

SOCIAL HISTORIES OF MEDICINE

Series Editors
David Cantor, Anne Hanley and Elaine Leong

Editorial Board
Diego Armus, Swarthmore College, PA, USA
Rana Hogarth, University of Illinois, Urbana-Champaign, USA
Angela Ki Che Leung, University of Hong Kong, China
Ian Miller, Ulster University, Northern Ireland

Social Histories of Medicine is concerned with all aspects of health, illness and medicine, from prehistory to the present, in every part of the world. The series covers the circumstances that promote health or illness, the ways in which people experience and explain such conditions, and what, practically, they do about them. Practitioners of all approaches to health and healing come within its scope, as do their ideas, beliefs, and practices, and the social, economic and cultural contexts in which they operate. Methodologically, the series welcomes relevant studies in social, economic, cultural, and intellectual history, as well as approaches derived from other disciplines in the arts, sciences, social sciences and humanities. The series is a collaboration between Manchester University Press and the Society for the Social History of Medicine.

To buy or to find out more about the books currently available in this series, please go to: https://manchesteruniversitypress.co.uk/series/social-histories-of-medicine/

Technology, health, and the patient consumer in the twentieth century

Edited by

Rachel Elder and Thomas Schlich

MANCHESTER UNIVERSITY PRESS

Published by Manchester University Press
Oxford Road, Manchester, M13 9PL

www.manchesteruniversitypress.co.uk

British Library Cataloguing-in-Publication Data
A catalogue record for this book is available from the British
Library

ISBN 978 1 5261 7114 6 hardback

First published 2025

Typeset by Newgen Publishing UK

Contents

Contents

List of figures

List of contributors

Fabiola Creed is a Postdoctoral Research Fellow in the School of Social & Political Sciences at the University of Glasgow.

Grazia De Michele is Gerald and Robin Silk Fellow at the Herbert D. Katz Center for Advanced Judaic Studies at the University of Pennsylvania.

Rachel Elder is Research Associate in the Department of Social Studies of Medicine at McGill University.

Vivien Hamilton is Professor of the History of Science at Harvey Mudd College.

Antoine Lentacker is Assistant Professor in the Department of History at the University of California, Riverside.

Richard M. Mizelle, Jr is Associate Professor in the Department of History at the University of Houston.

Christopher M. Rudeen is a lecturer in history at Smith College.

Thomas Schlich is James McGill Professor in the History of Medicine and Department Chair in the Department of Social Studies of Medicine at McGill University.

Cynthia L. Tang is Intermediate Specialist, Business Operations/ Strategic Initiatives at the Tefler School of Management at the University of Ottawa.

Sharra Vostral is Professor of Instruction in the Department of Communication Studies at Northwestern University, where she also serves as Assistant Dean for Research in the School of Communication.

Acknowledgements

This project is the culmination of a two-day conference co-hosted by the Department of Social Studies of Medicine at McGill University and the Jewish General Hospital in Montreal, Quebec, Canada, in 2021. We would like to thank all participants of the conference, including organizers, session chairs, research assistants, and those whose chapters are featured in this book. In addition, we would like to extend a special thanks to Dr Lawrence Rosenberg, Barbara Reney, Charlotte Maertens, David Cantor, Whitney Laemmli, our external reviewers, and the editorial team at Manchester University Press. A big thank you goes to Sophia Motluk for her excellent work and invaluable help in preparing the manuscript for publication. No portion of this project would have been possible without the generous funding of the Canadian Institutes of Health Research (CIHR) and the Social Sciences and Humanities Research Council of Canada (SSHRC).

Introduction: Technology and health in the age of the patient consumer

Thomas Schlich and Rachel Elder

Today, in contexts ranging from telehealth services to the health insurance landscape, 'patient' and 'consumer' have become almost interchangeable terms. At the same time, a growing array of technologies continues to expand the options available for people who interact with modern health systems, increasing their access to different kinds of information and care. Twenty-first-century patients are as likely to find medical advice through apps on their phones, buy health products to use at home, and connect with other patients online as they are to interface directly with physicians or go to a clinic. These developments are a consequence of the rapid expansion of consumerism and a broad array of informational, medical, and health technologies as two characteristic phenomena in the history of modern medicine and healthcare. This evolution has reshaped what it means to be a patient in recent history, and yet it is rarely explored by scholars. The relationship between technology and patients' roles as consumers is thus a promising field of inquiry and the subject of this book.[1]

Historians have identified a variety of origins for patient consumerism. Alex Mold, for example, sees it as a direct outgrowth of patient groups that formed in the United Kingdom and elsewhere during the third quarter of the twentieth century, articulating new expectations for patients' rights, access to information, and discernment in matters of one's own health.[2] Studying the United States, Nancy Tomes instead points to structural changes in the business of medicine starting in the late nineteenth century. Though Tomes' definition similarly encompasses grassroots health movements and consumer watchdog groups that would appear by the 1960s – as well as the more neoliberal connotations of patient consumerism that had emerged by the 1980s – her emphasis is on broad, long-range factors such as medical regulation and advertising.[3] There are consequently many associations with the idea of the patient consumer. The term can signify one's role, such as being an informed shopper or an activist, or point to the considerable inequities produced by capitalist systems of care. Its history implicates a wide cast of players, including medical professionals, drug manufacturers, health insurance companies, and the media. In all instances, elements of consumerism are linked with the roles of sick and healthy people, invoking ideals of individualism, choice, convenience, and control – aspirations that are embodied, too, in the technologies that increasingly permeate our health experiences.

The chapters ahead centre on understanding the interplay between some of these technologies and patient consumers in the United States and the United Kingdom. We focus on the early twentieth century to the present since this period witnessed a marked increase in the number and dissemination of technologies in and around medicine, including not only those located in hospitals, such as ventilators, fMRI (functional magnetic resonance imaging) machines, and intensive care units, but also home blood pressure-monitoring devices, assistive walkers, and health supplements.[4] It was also the period in which the 'patient consumer' emerged.[5] Collectively, the chapters illuminate a range of questions concerning the impact of these developments: do technologies improve, deteriorate, complicate patients' experiences? Do they increase or limit their access to care or multiply patients' medical and personal choices? How have consumerist-driven ways of being a patient changed the nature of patienthood, and what role have factors such as race, class, gender, and geography played in shaping these stories? Overall, the different chapters highlight the ambiguity of this history. On the one hand, the uptake of technologies by patient consumers can popularize new treatments and channels of information. It can just as easily, however, deepen existing disparities and cause patients to feel alienated from their health management. As historians, we want to examine such instances with case-by-case specificity. Ultimately, we show the extent to which technology and the patient consumer have remained forces that together shape the nature and experience of medical care.

Histories of patients and technology

Ever since historians Susan Reverby, David Rosner, and Roy Porter called for the move 'beyond great doctors' and a 'medical history from below', the search for the patient's voice has been a priority of the social history of medicine.[6] This has included efforts to understand patients' experiences and the extent of their power in determining the care they receive. Still, such histories are not straightforward. Scholars such as Nicholas Jewson and Michel Foucault claimed from the outset that the patient's voice was effaced around 1800 with the rise of clinical medicine in hospital settings.[7] Seen from this perspective, the very category of 'patient' in modern medicine is a construct of the 'clinical gaze'; patients exist in their modern role only from the doctor's perspective within a de-individualizing approach to medicine. The constructed character of patient identity thus poses a problem for historians when they try to capture the patient's perspective, as Flurin Condrau and Michael Brown point out.[8] Some have therefore resolved that any 'master narrative of a patient's history' should be replaced with 'carefully contextualised analyses of "patients"', considering that the ability to recover the patient's voice and even the definition of the term 'patient history' varies considerably.[9]

These obstacles to capturing patients in history are further complicated by the fact that the designation of 'patient' itself carries historical connotations of passivity and subordination.[10] Such a view, however, only partially covers the roles available. Many sick people in Victorian Britain did not, as Anne Hanley and Jessica Meyer recently noted, see themselves as patients, but as actively engaged 'health users' with certain degrees of power.[11] In the broader historiography of medicine, such agency is most often conceptualized within the framework of the 'medical marketplace', a concept rooted in the history of Early Modern Europe before the rise of professionalized medicine in the nineteenth century when practitioners competed for customers in an open marketplace, selling and advertising their services. 'Patients' in such settings had 'relative freedom to choose the medical practitioners they liked'; in short, 'they were medically promiscuous'.[12] Of course, such freedom, as other historians note, had its limits. One's ability to choose was always embedded in larger social, economic, and political structures. Such factors continue to shape care and its accessibility.[13]

The idea of a marketplace nevertheless lends itself to a characterization of patients as consumers, a concept that has continued to resonate in more recent periods, too. Though neither uses the language of a medical marketplace, both Nancy Tomes and Alex Mold have charted a growing conflation of the patient role with consumer behaviours over the course of the twentieth century, which produced what each calls the 'patient consumer'. According to their findings, transformations in the United States' and the United Kingdom's differing health systems, as well as the rise of an increasingly consumerist society, redefined how patients sought medical information and treatment, related to their physicians, and navigated the changing spaces of medical care.[14] Conceptualizing patients in this way also allows for more patient-oriented histories. Among other things, it expands the focus beyond medical institutions to other factors that shape patients' habits. As Tomes notes of the American context, '[t]hinking of patients as consumers suggests ways to connect a multiplicity of individual decisions that Americans have made about their health and health care ... with the evolution of twentieth-century health care institutions'.[15]

In looking at the patient consumer, we refer to this history of seeming autonomy and conditional choice within the context of larger market forces and the institutions of and around medicine. While some draw distinctions between 'patient consumer' and the somewhat broader category of 'health consumer', chapters in this volume highlight the ways that these categories can overlap. To consider patients as 'health users', moreover – 'user' being a particularly salient category in the history of technology – also makes it plausible to include what Michael Worboys has termed 'non-patients' – those sufferers who are not consulting a doctor.[16] However, we can go even further and broaden the definition of 'healthcare' as a dynamic and contested entity.[17] Processes of medicalization, for example, have long extended the boundaries of healthcare to include previously non-medical domains, many of which people have again begun to navigate without formal medical institutions or experts. A classic example is the reproductive sphere of contraception and birth, though one can imagine several others, including some of those explored in this volume.[18] Ultimately, taking a wide view of the patient consumer allows us to capture aspects of multiple health and illness experiences, including those unfolding beyond formal healthcare structures. Such an approach also encompasses a fuller range of behaviours and motivations suggested by the term 'consumer' by including, for example, shopping and pleasure-driven phenomena emerging from modern society's 'preoccupation with fulfilling, healthy lifestyles'.[19]

Yet until now, efforts to analyse the phenomenon of the patient consumer within the historical literature have had little to say about the role of technology. This remains a notable omission given that, historically, medicine, healthcare, and patients' own practices have been bound up with technologies. Though it has always been a part of medicine, historians, in fact, argue that technology first became a central, even inescapable, feature of healthcare in the early twentieth century, which further transformed the primary sites of care from home to hospital.[20] Others point to various medical technologies circulating beyond hospital contexts over the course of the same period, noting the ways in which they similarly transformed patients' experiences.[21] Feminist historians of technology, more generally, were among the first to emphasize questions of consumption, including mass production, distribution, and consumer markets as an important dimension of technology.[22] Where, then, does technology feature in the history of patient consumers? And what, moreover, counts as technology in medicine or indeed a 'medical technology'?

The term 'medical technology' tends to conjure images of large, complex machines located in hospitals, perhaps connected to patients, yet far removed from their sphere of knowledge or influence. A variety of research projects, however, have shown the category to be far more capacious, including everything from patient-operated devices and pharmaceuticals to the information systems that keep clinics and their personnel functioning. Indeed, as proposed by scholars in science and technology studies, the word 'technology' itself has three layers of meaning.[23] First, there is the physical level. With a hospital-based example such as surgery, this might involve the instruments and the operating room. Second, technology can be seen as an activity or a means to accomplish a specific goal. This would refer to what the surgical team does in the operative procedure. Finally, technology is what people know. Having and using instruments is not enough for a successful surgery; those involved also need to understand how to apply the tools and techniques within their sphere of activity. Technology, in other words, is not only what is tangible; it encompasses relevant knowledge and skills.[24] Questions of use are similarly complex. The primary user of surgical technology, for instance, is the surgeon, but the surgeon applies the technology to another participant in the setting – a patient – who in a way becomes the end-user of the technology and, in certain instances, an active participant in the employment of new tools and techniques.

In this volume, we consider an array of technologies meant to help patients, support health, and mitigate or prevent illness. This definition includes various therapeutic technologies (whether pills, processes, or products), informational technologies that provide patients with knowledge, and opportunities for dialogue, as well as health devices and merchandise found in both domestic and clinical settings. Technology, in other words, exists in all spheres of health and medicine. It is not, as often presumed, necessarily new, complex, or superior to other therapeutic or diagnostic options. Nor is it only used by 'experts'.[25] Technologies are not the straightforward fulfilment of medical needs in simple and unmediated ways. Instead, they are influenced by many contextual factors. As such, researchers have looked not only to the invention or implementation of prominent technologies; they have also examined processes by which they become useful, or in some cases obsolete.[26] In the words of Jennifer Stanton, historians must treat medical technology as yet 'another element in the social, political, and economic history of medicine relating it to professionalization, changing notions about disease, and the organization of medical care'.[27]

It therefore continues to be worth investigating how specific medical technologies exist in relation to multiple, heterogeneous agents and structures, including doctors, manufacturers, institutions, regulators, and indeed patient consumers. To do so, one must consider how such technologies operate in a variety of contexts, and how they in turn shape and are shaped by them.[28] Such analysis further undermines the misleading idea that technology is some kind of autonomous force that by itself determines society or the way we live.[29] Pregnancy home testing, for example, was not enabled by the creation of new technologies alone. It also depended on vast cultural changes to normalize the method, including those exemplified by feminism and consumerism.[30] Focusing on a broader array of 'users' also shifts the balance away from doctors and inventors, and emphasizes how diverse actors, including patients and other healthcare professionals, have given technology form and meaning.[31] Margarete Sandelowski, for instance, places nurses at the centre of complex negotiations surrounding many new medical technologies of the twentieth century, and indeed older ones, in ways that were consequential for patients and their health.[32] Joel Howell similarly shows how X-rays and routine blood tests, as well as inconspicuous office management and accounting systems, significantly changed not just medical practice but also patients' experiences in the early twentieth century.[33]

Still, what does technology give, or take away from, patients, if anything? What has been the impact and role of technology in medicine as it relates to the patient consumer, specifically? Simple narratives of progress are of limited usefulness to answer such questions. In such stories, a particular medical device may emerge to serve patients in ways that are unequivocally good, just, and perhaps even inevitable.[34] These narratives can also overlook patients. However, perhaps even more common in the history of medicine is a tradition of blaming technology for distancing physicians from their patients, a development that supposedly happened in conjunction with the devaluation of personal accounts and the limiting of patients' power. In his classic 1978 text, *Medicine and the Reign of Technology*, Stanley Reiser, for instance, traced the downside of such technological advances 'to the sick patient, to the physician as clinician, and to society'.[35] He predicted a future that was 'extremely disquieting' and warned of the possibility that physicians could adopt a 'fundamentally mechanical view of human beings'. The 'high price for the use of technology', he said, might be 'impersonal medical care'.[36]

Early on, Roy Porter claimed that patient autonomy and choice had been much stronger in the marketplace situation of the Early Modern period, a time before doctors could point to 'gleaming technology and miracle cures to bolster their mystique'.[37] If at all, this is only partially true. As this volume highlights, technology has not necessarily limited patient choices, but often enhanced them. Patients, moreover, have never been merely passive recipients of technology: they are quite often users and sometimes even innovators. However, the opposite is not true either. Technology has never automatically turned patients into actively engaged health consumers. Instead, historians need to analyse how patients and technologies operate together. Hanley and Meyer, for instance, describe how patients and technologies have together been instrumental in a process of 'user driven co-production', including in the 'design and redesign of health networks, services and technologies'.[38] Patients' options have been increased even more by the spread of inexpensive and readily available devices and treatments in domestic and commercial settings, allowing some to bypass hospitals and doctors' offices altogether.[39] In this context, patients can take on the role of consumer, a role that comes with specific constellations of power. Patient consumers acquire, at least theoretically, a particular range of rights and the possibility to 'vote with their wallets', dependent, of course, on the resources they have at their disposal.[40] Even those ambivalent about technology have interpreted the consumer health movement as an avenue for counterbalancing the potentially harmful effects of medical technology.[41]

Technology and patient consumers

The chapters in this volume elucidate these issues in finer detail by bringing together two separate strains of research in the history of medicine and technology. Our emphasis on the twentieth- and twenty-first-century United States and United Kingdom in part reflects a growing concentration of technological innovation and consumption within these regions, in which an accelerated pace of development stoked confidence in technology as a progressive force, and medical technologies themselves began to proliferate at an exceptional rate.[42] There is still, of course, much to explore by historians and social scientists to explain how such stories have unfolded in other contexts, where the nature of the relationships between technology and patients may, or may not, differ in meaningful ways.

The book is organized into three parts: I) New technologies and patient markets, II) Informed patients and patient information, and III) Co-opting disease, promoting prevention and healing.

Part I looks at how specific groups of patients influenced, or were thought to influence, emerging technologies in different areas of medicine from the 1920s to the 2020s. It also considers those patient consumers who failed to access new and often essential medical technologies, whether because of circumstance, structural inequalities, or, relatedly, because they were never prioritized users. Part II centres on questions of who owns medical and patient information, who is a reliable narrator and source of knowledge, and by which channels such information should flow or be controlled. In this part, internet databases and consumer records are the technologies in question, but also 'hygiene' products and psychopharmaceuticals which were at the centre of consumer and patient injury claims. Part III then shifts the focus to understanding how a wide variety of ills, ranging from cancers and precancers to gun violence, have inspired products claiming to prevent and heal them. In each of these cases, a technology of a kind becomes an imagined solution to a medical or medicalized 'problem', simultaneously blurring the boundaries of what constitutes a 'medical' versus a 'health' product, and when patients are simply patients, consumers, or some combination of the two.

One of the new technologies discussed in Part I is X-rays. While X-rays have been well researched in medicine and popular culture, their use in dentistry has not been examined. Looking at how this cutting-edge technology was first marketed to American dentists in the 1920s, Vivien Hamilton shows the ways in which White, well to-do patients became the promised clientele. She further discusses technology in its symbolic dimension as a factor for enhancing patients' trust. Paradoxically, however, the targeted dentists were reassured that they did not need special knowledge or skill to use the machines marketed to them. Another technology explored in this part is haemodialysis for kidney failure. Richard M. Mizelle, Jr analyses patients' access to this life-saving technology and how inequalities along lines of race, class, and disability continue to be exploited by profit-seeking companies with monopolies in certain regions of the United States. Specifically, he shows that African Americans in the South have been especially disenfranchised by the business models of the major providers, but also by haemodialysis' longer history and legacy of racism. In her chapter on minimally invasive surgery, Cynthia L. Tang traces how this new surgical technique was conceptualized and marketed across the United States in the 1990s. Chiefly, she challenges the often-repeated assumption that patients 'demanded' its introduction, thus fostering its rapid uptake and dissemination as the leading treatment for conditions such as gallstones. Instead, she shows that patient demand was not spontaneous but rather elicited by companies manufacturing the necessary equipment and by those surgeons who initially offered the new technique.

Part II on patient information begins by considering the shifting identities of women suffering from tampon-induced toxic shock syndrome as consumers, patients, and informants in the United States. In Sharra Vostral's chapter, menstruating women are given such varying roles in the context of controversies over the attribution of responsibilities for the disastrous side effects of a particular kind of body-care technology, a gendered technology, she shows, that straddles both medical and non-medical domains and contains numerous assumptions about women's need for privacy and 'protection'. In the subsequent chapter, Antoine Lentacker follows patients' voices on the internet and analyses how social media has challenged both the nature and place of patient information in Britain over the past thirty years. Drawing on the example of RxISK.org, he considers patients' own accounts concerning the safety of medications, in this case antidepressants, in view of long-standing hierarchies of scientific evidence within drug research. In this more recent case, patients' accounts of their side effects, long dismissed as anecdotal and anomalous, find a place as valid data to be gathered and shared, making patients themselves producers in a therapeutic knowledge economy.

Part III begins its discussion of technologies of prevention and healing with a chapter on consumer tanning practices in the second half of the twentieth century. Fabiola Creed shows that purportedly 'safe' tanning products in the United Kingdom, both physical and chemical ones, increasingly competed in a market saturated with claims about health and cancer prevention, yet ultimately capitalized upon mainstream beauty standards that continued to valorize white, tanned skin. Disease management and prevention are also at the centre of Grazia De Michele's contribution. Her chapter considers the role of organized patient movements in identifying and publicizing the harms associated with tamoxifen, a preventative pharmaceutical for breast cancer that turned out to contain its own risks. She places this case within broader controversies about cancer prevention in the 1990s, as well as a longer history of the hormones prescribed to women, showing the ways in which patient consumerism has continued to be defined by instances of patient resistance and activism. Finally, Christopher M. Rudeen conceptualizes commemorative T-shirts as a technology of group and community healing in reaction to racialized police violence across the United States. He discusses this technology in a complex field of tension between emancipatory self-treatment and commercialism from the 1990s to the present, noting the ways in which the ownership and impact of such articles remain difficult to disaggregate from their wider networks of consumption and inflicted harm.

As these examples show, the history of technology and patient consumers is a dynamic one with wide variations across time and place that continue even today. During the most recent COVID-19 pandemic, for instance, patients and non-patients took on new roles and identities. Many consumed rapid home-testing kits, masks, and vaccines in new constellations. They also parsed information through an unparalleled number of social media and news sources. Independent of the pandemic, there are longer-term developments that have moved medicine away from a professional monopoly towards technologically mediated roles for patient and health consumers. Telemedicine, the internet, and genetic testing are but a few examples.[43] According to some, medicine is undergoing its own process of 'uberization' or disintermediation – cutting out the middleman. In some contexts, doctors are displaced from their traditional position as an intermediary or gatekeeper, reviving perennial anxieties about the degradation of doctor–patient relationships through new technology, and, equally, utopian visions of self-reliance and care. Looking at histories of patient consumers provides a longer perspective on this ongoing revolution in medicine.

Among the lessons that may already be drawn from this history, perhaps most notable is the degree to which one's role as patient consumer or access to certain technologies is governed by social factors. As chapters by Richard M. Mizelle, Jr and Christopher M. Rudeen show, the technologies one can access, or that one must create to heal, are inextricable from systemic racism, socioeconomic inequality, and ableism. Similar insights about gender, and particularly the supposed vulnerability and vanity of female patient consumers, also emerge from the chapters of Vostral, De Michele, Hamilton, Creed, and Tang in different ways. Added to this, we see how patient consumers often assume additional roles on account of their social positions, whether that of informant, passive patient, or non-medical health consumer. Another key lesson is that the boundaries between patients and consumers – and indeed the quality of being a patient consumer – are not absolute; nor are the dynamics unidirectional. There are many instances in which technologies help turn regular consumers into patients, as the example of tampon-related injury shows. Likewise, technology sometimes enables patients to be seen in the ways that more accurately reflect their own experiences, as revealed by the example of patient side effect databases. Technology, in other words, neither turns patients into 'patient consumers' or 'health consumers', nor does it invariably expand or improve their chances of favourable treatment. Ultimately, being a patient consumer of technology is neither a minefield of potential harm nor a marketplace of unfettered choice and exploration as often suggested. Case studies such as the ones in this volume help make such complexity and ambiguity visible. They can thus enable a more varied and realistic picture of the patient consumer and the risks and opportunities associated with this new and ever-changing role.

Notes

1 Studies of both technology and consumerism point to the early twentieth century as a period in which the daily operation and appearance of medicine changed significantly. See Joel D. Howell, *Technology in the Hospital: Transforming Patient Care in the Early Twentieth Century* (Baltimore, MD: Johns Hopkins University Press, 1995); Nancy Tomes, 'Merchants of Health: Medicine and Consumer Culture in the United States, 1900–1940.' *Journal of American History* 88, no. 2 (2001): 519–47.

2 Alex Mold, *Making British Patients into Consumers: Patient Organizations and Health Consumerism in Britain*. Manchester: Manchester University Press, 2015; Alex Mold, 'Making the Patient-Consumer in Margaret Thatcher's Britain.' *Historical Journal* 54, no. 2 (2011): 509–28. See also Glen O'Hare, 'The Complexities

of "Consumerism": Choice, Collectivism and Participation within Britain's National Health Service, c. 1961–c. 1979.' *Social History of Medicine* 26, no. 2 (2013): 288–304.

3 Nancy Tomes, *Remaking the American Patient: How Madison Avenue and Modern Medicine Turned Patients into Consumers* (Chapel Hill, NC: University of North Carolina Press, 2016).

4 On technology in medicine in the twentieth century, see Howell, *Technology in the Hospital*; Margarete Sandelowski, *Devices and Desires: Gender, Technology, and American Nursing* (Chapel Hill, NC: University of North Carolina Press, 2000); Keith Wailoo, *Drawing Blood: Technology and Disease Identity in Twentieth-Century America* (Baltimore, MD: Johns Hopkins University Press, 1999); Jeffrey Baker, *The Machine in the Nursery: Incubator Technology and the Origins of Newborn Intensive Care* (Baltimore, MD: Johns Hopkins University Press, 1996).

5 Tomes, *Remaking the American Patient*; Mold, *Making British Patients into Consumers*.

6 Susan Reverby and David Rosner, 'Beyond "the Great Doctors"', in *Health Care in America: Essays in Social History*, edited by Susan Reverby and David Rosner, pp. 3–16 (Philadelphia, PA: Temple University Press, 1979); Roy Porter, 'The Patient's View: Doing Medical History from Below.' *Theory and Society* 14 (1985): 175–98.

7 Michel Foucault, *The Birth of the Clinic* (New York: Pantheon, 1973); N.D. Jewson, 'The Disappearance of the Sick-Man from Medical Cosmology.' *Sociology* 10 (1976): 369–85.

8 Michael Brown, *Emotions and Surgery in Britain, 1793–1912* (Cambridge: Cambridge University Press, 2023), pp. 110–14; Flurin Condrau, 'The Patient's View Meets the Clinical Gaze.' *Social History of Medicine* 20, no. 3 (2007): 525–40.

9 Condrau, 'The Patient's View Meets the Clinical Gaze', 536.

10 Anne Hanley and Jessica Meyer, 'Introduction: Searching for the Patient', in *Patient Voices in Britain, 1840–1948*, edited by Anne Hanley and Jessica Meyer, pp. 1–29 (Manchester: Manchester University Press, 2021), p. 4.

11 Hanley and Meyer, 'Searching for the Patient', p. 5.

12 Mark Jenner and Patrick Wallis, *Medicine and the Market in England and Its Colonies, c. 1450–c. 1850* (New York: Palgrave Macmillan, 2007), p. 2.

13 Hanley and Meyer, 'Searching for the Patient', p. 7.

14 While the literature on medical consumerism is relatively robust, 'patient consumerism' is still quite a small field of study. See Mold, *Making British Patients into Consumers*; Tomes, *Remaking the*

American Patient; O'Hare, 'The Complexities of "Consumerism"', and, more recently, Hanley and Meyer, *Patient Voices*.

15 Tomes, 'Merchants of Health', 523.

16 Michael Worboys, 'The Non-Patient's View', in Hanley and Meyer, *Patient Voices*, pp. 33–60.

17 Roberta Bivins, Hilary Marland, and Nancy Tomes, 'Histories of Medicine in the Household: Recovering Practice and "Reception".' *Social History of Medicine* 29 (2016): 669–75.

18 See, for example, Jesse Olszynko-Gryn, *A Woman's Right to Know: Pregnancy Testing in Twentieth-Century Britain* (Cambridge, MA: MIT Press, 2023), p. 372. The classic reference is Judith Walzer Leavitt, *Brought to Bed: Childbearing in America, 1750 to 1950* (New York: Oxford University Press, 1986).

19 Hanley and Meyer, 'Searching for the Patient', p. 4.

20 Howell, *Technology in the Hospital*.

21 For example, see Carolyn Thomas de la Pena, *The Body Electric: How Strange Machines Built the Modern American* (New York: New York University Press, 2005); Rima Apple, *Vitamania: Vitamins in American Culture* (New Brunswick, NJ: Rutgers University Press, 1996).

22 Ruth Schwartz Cowan, 'The Consumption Junction: A Proposal for Research Strategies in the Sociology of Technology', in *The Social Construction of Technological Systems: New Directions in the Sociology and History of Technology*, edited by Wiebe E. Bijker, Thomas P. Hughes, and Trevor Pinch, pp. 261–80 (Boston, MA: MIT Press, 1987); see also Roger Horowitz and Arwen Mohun (eds), *His and Hers: Gender, Consumption, and Technology* (Baltimore, MD: Johns Hopkins University Press, 1998); Ruth Schwartz Cowan, *More Work for Mother: The Ironies of Household Technologies from the Open Hearth to the Microwave* (New York: Basic Books, 1983).

23 Wiebe E. Bijker, Thomas P. Hughes, and Trevor Pinch, 'General Introduction', in Bijker, Hughes, and Pinch, *The Social Construction of Technological Systems*, pp. 1–15, see p. 4; Howell, *Technology in the Hospital*, p. 8. For the use of this concept in surgery, see Thomas Schlich, *Surgery, Science and Industry: A Revolution in Fracture Care, 1950s–1990s* (Basingstoke: Palgrave Macmillan, 2002).

24 Sandelowski, *Devices and Desires*, p. 26.

25 David Edgerton, *The Shock of the Old: Technology and Global History since 1900* (New York: Oxford University Press, 2007); Sandelowski, *Devices and Desires*.

26 Thomas Schlich and Christopher Crenner, 'Technological Change in Surgery: An Introductory Essay', in *Technological Change in Modern Surgery: Historical Perspectives on Innovation*, edited by Thomas

Schlich and Christopher Crenner, pp. 1–20 (Rochester, NY: University of Rochester Press, 2017); John V. Pickstone, 'Introduction', in *Medical Innovations in Historical Perspective*, edited by John V. Pickstone, pp. 1–16 (Basingstoke: Macmillan, 1992), p. 16.

27 Jennifer Stanton, 'Making Sense of Technologies in Medicine.' *Social History of Medicine* 12 (1999): 437–48, 438.

28 Trevor J. Pinch and Wiebe E. Bijker, 'The Social Construction of Facts and Artifacts: Or How the Sociology of Science and the Sociology of Technology Might Benefit Each Other', in Bijker, Hughes, and Pinch, *The Social Construction of Technological Systems*, pp. 17–50, see pp. 22–4; Stanton, 'Making Sense'; also, Schlich, *Surgery, Science and Industry*, p. 239; Howell, *Technology in the Hospital*, p. 11.

29 Leo Marx and Merritt Roe Smith (eds), *Does Technology Drive History? The Dilemma of Technological Determinism* (Cambridge, MA: MIT Press, 1994); Howell, *Technology in the Hospital*, pp. 227–8.

30 Olszynko-Gryn, *A Woman's Right*, pp. 371–82.

31 Nelly Oudshorn and Trevor Pinch (eds), *How Users Matter: The Co-Construction of Users and Technology* (Boston, MA: MIT Press, 2005); Edgerton, *The Shock of the Old*, p. 8.

32 Sandelowski, *Devices and Desires*.

33 Howell, *Technology in the Hospital*, pp. 1–29.

34 Schlich and Crenner, 'Technological Change'.

35 Stanley Joel Reiser, *Medicine and the Reign of Technology* (Cambridge, New York, Melbourne: Cambridge University Press, 1978), p. ix. See also Stanley Joel Reiser, *Technological Medicine: The Changing World of Doctors and Patients* (New York: Cambridge University Press, 2009), p. 188.

36 Reiser, *Technological Medicine*, pp. 229–31.

37 Porter, 'The Patient's View', 189.

38 Hanley and Meyer, 'Searching for the Patient', p. 5.

39 Sandelowski, *Devices and Desires*; see, for example, Lawrence Rosenberg, *Patients Matter Most: How Healthcare Is Becoming Personal Again* (Charleston, SC: Forbes, 2023).

40 For example, Olszynko-Gryn, *A Woman's Right*, p. 372.

41 Reiser, *Technological Medicine*, p. 188.

42 See, for example, Allan M. Brandt and Martha Gardner, 'The Golden Age of Medicine?', in *Companion to Medicine in the Twentieth Century*, edited by Roger Cooter and John Pickstone, pp. 21–38 (London and New York: Routledge, 2002); Bert Hansen, *Picturing Medical Progress from Pasteur to Polio: A History of Mass Media Images and Popular Attitudes in America* (New Brunswick,

NJ: Rutgers University Press, 2009). Thomas Schlich, ' "One and the Same the World Over" – The International Culture of Surgical Exchange in an Age of Globalization, 1870–1914.' *Journal of the History of Medicine and Allied Sciences* 71 (2016): 247–70.
43 See, for example, Rosenberg, *Patients Matter Most*. On telemedicine, see Jeremy Greene, *The Doctor Who Wasn't There: Technology, History, and the Limits of Telehealth* (Chicago, IL: University of Chicago Press, 2022).

Bibliography

Apple, Rima. *Vitamania: Vitamins in American Culture*. New Brunswick, NJ: Rutgers University Press, 1996.

Baker, Jeffrey. *The Machine in the Nursery: Incubator Technology and the Origins of Newborn Intensive Care*. Baltimore, MD: Johns Hopkins University Press, 1996.

Bijker, Wiebe E., Thomas P. Hughes, and Trevor Pinch. 'General Introduction', in *The Social Construction of Technological Systems: New Directions in the Sociology and History of Technology*, edited by Wiebe E. Bijker, Thomas P. Hughes, and Trevor Pinch, pp. 1–15. Boston, MA: MIT Press, 1987.

Bivins, Roberta, Hilary Marland, and Nancy Tomes. 'Histories of Medicine in the Household: Recovering Practice and "Reception".' *Social History of Medicine* 29 (2016): 669–75.

Brandt, Allan M., and Martha Gardner. 'The Golden Age of Medicine?', in *Companion to Medicine in the Twentieth Century*, edited by Roger Cooter and John Pickstone, pp. 21–38. London and New York: Routledge, 2002.

Brown, Michael. *Emotions and Surgery in Britain, 1793–1912*. Cambridge: Cambridge University Press, 2023.

Condrau, Flurin. 'The Patient's View Meets the Clinical Gaze.' *Social History of Medicine* 20, no. 3 (2007): 525–40.

Edgerton, David. *The Shock of the Old: Technology and Global History since 1900*. New York: Oxford University Press, 2007.

Foucault, Michel. *The Birth of the Clinic*. New York: Pantheon, 1973.

Greene, Jeremy. *The Doctor Who Wasn't There: Technology, History, and the Limits of Telehealth*. Chicago, IL: University of Chicago Press, 2022.

Hanley, Anne, and Jessica Meyer. 'Introduction: Searching for the Patient', in *Patient Voices in Britain, 1840–1948*, edited by Anne Hanley and Jessica Meyer, pp. 1–29. Manchester: Manchester University Press, 2021.

Hansen, Bert. *Picturing Medical Progress from Pasteur to Polio: A History of Mass Media Images and Popular Attitudes in America*. New Brunswick, NJ: Rutgers University Press, 2009.

Horowitz, Roger, and Mohun Arwen (eds). *His and Hers: Gender, Consumption, and Technology*. Baltimore, MD: Johns Hopkins University Press, 1998.

Howell, Joel D. *Technology in the Hospital: Transforming Patient Care in the Early Twentieth Century*. Baltimore, MD: Johns Hopkins University Press, 1995.

Jenner, Mark, and Patrick Wallis. *Medicine and the Market in England and Its Colonies, c. 1450–c. 1850*. New York: Palgrave Macmillan, 2007.

Jewson, N.D. 'The Disappearance of the Sick-Man from Medical Cosmology.' *Sociology* 10 (1976): 369–85.

Marx, Leo, and Merritt Roe Smith (eds). *Does Technology Drive History? The Dilemma of Technological Determinism*. Cambridge, MA: MIT Press, 1994.

Mold, Alex. *Making British Patients into Consumers: Patient Organizations and Health Consumerism in Britain*. Manchester: Manchester University Press, 2015.

Mold, Alex. 'Making the Patient-Consumer in Margaret Thatcher's Britain.' *Historical Journal* 54, no. 2 (2011): 509–28.

O'Hare, Glen. 'The Complexities of "Consumerism": Choice, Collectivism and Participation within Britain's National Health Service, c. 1961–c. 1979.' *Social History of Medicine* 26, no. 2 (2013): 288–304.

Olszynko-Gryn, Jesse. *A Woman's Right to Know: Pregnancy Testing in Twentieth-Century Britain*. Cambridge, MA: MIT Press, 2023.

Oudshorn, Nelly, and Trevor Pinch (eds). *How Users Matter: The Co-Construction of Users and Technology*. Boston, MA: MIT Press, 2005.

Pickstone, John V. 'Introduction', in *Medical Innovations in Historical Perspective*, edited by John V. Pickstone, pp. 1–16. Basingstoke: Macmillan, 1992.

Pinch, Trevor J., and Wiebe E. Bijker. 'The Social Construction of Facts and Artifacts: Or How the Sociology of Science and the Sociology of Technology Might Benefit Each Other', in *The Social Construction of Technological Systems: New Directions in the Sociology and History of Technology*, edited by Wiebe E. Bijker, Thomas P. Hughes, and Trevor Pinch, pp. 17–50. Boston, MA: MIT Press, 1987.

Porter, Roy. 'The Patient's View: Doing Medical History from Below.' *Theory and Society* 14 (1985): 175–98.

Reiser, Stanley Joel. *Medicine and the Reign of Technology*. Cambridge, New York, Melbourne: Cambridge University Press, 1978.

Reiser, Stanley Joel. *Technological Medicine: The Changing World of Doctors and Patients*. New York: Cambridge University Press, 2009.

Reverby, Susan, and David Rosner. 'Beyond "the Great Doctors"', in *Health Care in America: Essays in Social History*, edited by Susan Reverby and David Rosner, pp. 3–16. Philadelphia, PA: Temple University Press, 1979.

Rosenberg, Lawrence. *Patients Matter Most: How Healthcare Is Becoming Personal Again*. Charleston, SC: Forbes, 2023.

Sandelowski, Margarete. *Devices and Desires: Gender, Technology, and American Nursing*. Chapel Hill, NC: University of North Carolina Press, 2000.

Schlich, Thomas. '"One and the Same the World Over" – The International Culture of Surgical Exchange in an Age of Globalization, 1870–1914.' *Journal of the History of Medicine and Allied Sciences* 71 (2016): 247–70.

Schlich, Thomas. *Surgery, Science and Industry: A Revolution in Fracture Care, 1950s–1990s*. Basingstoke: Palgrave Macmillan, 2002.

Schlich, Thomas, and Christopher Crenner. 'Technological Change in Surgery: An Introductory Essay', in *Technological Change in Modern Surgery: Historical Perspectives on Innovation*, edited by Thomas Schlich and Christopher Crenner, pp. 1–20. Rochester, NY: University of Rochester Press, 2017.

Schwartz Cowan, Ruth. 'The Consumption Junction: A Proposal for Research Strategies in the Sociology of Technology', in *The Social Construction of Technological Systems: New Directions in the Sociology and History of Technology*, edited by Wiebe E. Bijker, Thomas P. Hughes, and Trevor Pinch, pp. 261–80. Boston, MA: MIT Press, 1987.

Schwartz Cowan, Ruth. *More Work for Mother: The Ironies of Household Technologies from the Open Hearth to the Microwave*. New York: Basic Books, 1983.

Stanton, Jennifer. 'Making Sense of Technologies in Medicine.' *Social History of Medicine* 12 (1999): 437–48.

Thomas de la Pena, Carolyn. *The Body Electric: How Strange Machines Built the Modern American*. New York: New York University Press, 2005.

Tomes, Nancy. 'Merchants of Health: Medicine and Consumer Culture in the United States, 1900–1940.' *Journal of American History* 88, no. 2 (2001): 519–47.

Tomes, Nancy. *Remaking the American Patient: How Madison Avenue and Modern Medicine Turned Patients into Consumers*. Chapel Hill, NC: University of North Carolina Press, 2016.

Wailoo, Keith. *Drawing Blood: Technology and Disease Identity in Twentieth-Century America*. Baltimore, MD: Johns Hopkins University Press, 1999.

Walzer Leavitt, Judith. *Brought to Bed: Childbearing in America, 1750 to 1950*. New York: Oxford University Press, 1986.

Worboys, Michael. 'The Non-Patient's View', in *Patient Voices in Britain, 1840–1948*, edited by Anne Hanley and Jessica Meyer, pp. 33–60. Manchester: Manchester University Press, 2021.

Part I

New technologies and patient markets

1

Dental X-rays and the imagined patient

Vivien Hamilton

By the 1920s, X-rays were a stable component of hospital medical care in the United States, used both as a diagnostic tool and as a therapeutic agent to treat skin conditions, cancers, and other tumours.[1] While the diagnostic benefits of X-rays seemed equally alluring to dentists, this technology took longer to become a standard feature of dental offices. Like doctors, some dentists started experimenting with X-rays right after their discovery in 1895.[2] Twenty-five years later, a few vocal advocates for dental X-rays continued to note the great promise of a technology that could reveal cavities and other infections, and assist in root canal treatment, oral surgeries, crowns, and bridgework.[3] Most dentists, however, still needed to be persuaded to purchase X-ray equipment. Advertising in the *Journal of the American Dental Association* in 1922, Victor X-Ray Company urged dentists: 'Don't Forget a Xmas Gift for Yourself', promising that an X-ray machine would be a 'lasting, useful … companion'.[4] X-ray equipment was portrayed as a bit of an indulgence, not yet a crucial component of dental care.

By the early 1930s, many practices had been hit hard by the Depression, and there was even more work to do to persuade American dentists to purchase an X-ray unit or upgrade their existing equipment. Equipment sold by Ritter X-Ray Company, a German organization with operations in New York, cost about $1,000 at the time (equivalent to $20,000 today).[5] This was on top of the $2,000–$3,000 that new dentists needed to spend to equip a small dental office.[6] Given this cost, it is not surprising that a survey of newly graduated dentists in 1930 found that only 46 per cent bought their own X-ray machine.[7] Established dentists wishing to purchase an X-ray machine might also need to contemplate a renovation of their existing space in order to create a new X-ray room, or at least a small dark room for developing the X-ray films.[8] With these costs and potential disruptions in mind, salespeople were told that 'the ability to size up and approach a dentist as a buyer is useful in determining how to open the interview – how to get under his skin, you might say'.[9] A Ritter training manual generated a list of twenty-five common objections that salespeople might hear from dentists unsure about buying their own X-ray machine, with the aim of coaching their employees on how to react. Salespeople were told to expect dentists to appeal to bad finances, to say that they preferred to wait to have cash to pay for the equipment, would rather spend money on a wedding, or were content to send X-ray work out to external labs. At a time when X-ray equipment was still seen as a luxury, dentists might simply say: 'I just don't want an X-ray machine'.[10] Salespeople were coached to make promises about the positive impression that X-ray equipment would make on paying customers. These customers would be newly attracted to the dentist's practice by this technology, willing to pay both for X-rays and for any dental procedures indicated by these images.

It is these imagined consumer patients that I am most interested in exploring in this chapter, not only to uncover their central role in X-ray marketing but also to understand how expectations about dental patients were reflected in the design of the technology itself. Focusing on marketing allows us to pay attention to what Ruth Schwartz Cowan calls 'the consumption junction, the place and time at which the consumer makes choices between competing technologies'.[11] For historians of technology, paying attention to consumer choice offers a particular kind of insight, complementary to studies focusing on innovation, technological development, and the day-to-day use of technologies.

The consumption junction is a crucial location for studying the diffusion of a technology, helping us understand which character-istics, or which technologies, were valued by consumers and why. In this story, there are two important groups of consumers: dentists buying and operating X-ray machines and patients purchasing dental care. In the language of the social construction of technol-ogy (SCOT), both are relevant social groups, yet with very different relationships to the technology.[12] Unlike different groups of bicycle users who valued different bicycle designs, dental patients were not directly choosing certain features of X-ray equipment. It was the dentist making the decision to buy a particular X-ray machine, or not, and so we would expect patients as consumers to have minimal influence over the design of X-ray technology. What I find in this case study, however, is that a particular kind of imagined patient *was* a central player in the consumption junction, invoked by X-ray companies hoping to encourage dentists to make a very expensive purchase. These ideal patients were crucial to the perceived promise of X-rays for dentists, and even more, they can help us make sense of some key features of dental X-ray equipment and practice.

In this chapter, I take a closer look at catalogues and pamphlets advertising X-ray technology to dentists, as well as training material for X-ray salespeople, in order to uncover the imagined patient reflected in this material. I focus primarily on trade literature produced by Ritter X-Ray Company, perhaps the largest manufacturer of X-ray equipment for dentists. Ritter launched an extensive advertising campaign targeting dentists in the 1930s, a time when X-ray technology was not yet a standard feature of all dental practices.[13] This advertising invoked ideal patients who were educated, demanding consumers insisting on modern dental techniques. X-ray technology was expected to function as a powerful symbol of modern, scientific dentistry for these patients. However, where the ads promised that patients would see the dentist as an expert creator of X-ray images, in reality, the equipment itself took many decisions out of the hands of dentists, who were assured that little skill or knowledge was actually necessary to operate the machines. Looking at the design of dental X-ray machines from this same period, it is clear that dentists did not want to have to exercise complex judgement in using the technology. Dentists purchasing this equipment were content to put their trust in the machines, with the expectation that simply producing an X-ray image would be a sufficient performance of scientific dentistry for their imagined patients.

Social advancement for (white) patients and dentists

Manufacturers promised that an X-ray machine would attract certain kinds of patients: patrons who desired the social uplift made possible by visibly healthy teeth, and who would choose a dental clinic based on their perception of the dentist's skill and education. The fact that Ritter could invoke such a patient speaks to the perceived success of public health initiatives in the preceding decades that had created a wider market for dental care in the United States. In the early twentieth century, very few Americans received dental care. One estimate in 1909 was that only 5–8 per cent of the population regularly visited a dentist.[14] In order to increase their reach, dentists regularly supported school programmes designed to teach American children about the importance of regular dental care, not just to address problems like cavities but to prevent those problems in the first place. One paediatric dentist in the 1920s recommended that children visit her office for monthly cleaning and polishing, though that schedule was clearly quite ambitious.[15] Statistics from 1929 showed that even in households making $2,000–$3,000 (more than the national average), only 21 per cent visited the dentist once per year.[16]

Dentists hoped that school outreach programmes would increase demand for dental services at a time when professional ethics discouraged them from advertising directly to patients. By teaching children about germs, and demonstrating the link between dental health and overall health, school programmes emphasized the benefits of regular check-ups and professional cleaning to prevent tooth decay. Dentists often expressed frustration that patients seemed to prefer extractions over other kinds of treatments, like root canals or fillings that could preserve a tooth. A key goal of these programmes was to instill in children a sense that teeth were valuable, beautiful, and worth saving, with the hope that they would brush their teeth more carefully, and also convince their parents to pay for dental care.[17]

In addition to these public health goals, dentists expected better-educated patients to demand better dentists, reducing the market for quackery. As one dentist hoped, 'A public thus educated will demand intelligent, capable, scientific dentists, and will also elevate the standard of the profession.'[18] While dental licencing requirements had been adopted by all states by the end of the nineteenth century, these requirements were initially fairly light. To obtain a licence, dentists needed to show that they had graduated from a dental school, but with the proliferation of proprietary schools in the late nineteenth century, curriculum and training in dental schools varied widely. Slowly, more and more states adopted a standard licencing exam. In 1926 the Carnegie-funded Gies Report functioned like a Flexner Report for dentistry, calling on universities to pay more attention to their dental schools and invest in a foundational science curriculum. The Report articulated a number of criticisms of dental education that had already been circulating, and by the time it was published, only three proprietary dental schools remained in the United States.[19] Still, standardization of dental education and licencing was slow across the country, and many dentists felt that they faced competition from colleagues with questionable credentials or training.

In this dental marketplace, equipment manufacturers emphasized that patients would form their strongest impression of a dentist based on the atmosphere and environment of the dental clinic. As an ad for S.S. White Operating Equipment notes, 'knowledge, skill and experience are too often judged last by patients. They estimate a man's ability first by his surroundings, then by his personality, his skills and training last. Fair or unfair, this method of mass approval is nevertheless a fact'.[20] An X-ray machine was a large, prominent feature of a dental office, sure to be noticed by patients. The British X-ray company Watson & Sons promised: 'The installation of X-ray equipment cannot but have a salutary effect on the mental impressions your patients have of your methods … [offering] a constant and effective advertisement, a silent creator of confidence and goodwill', even when not in operation.[21] At a time when advertising was professionally taboo for dentists, patients may not have known in advance that a particular dental clinic owned an X-ray machine. Historian Stine Grumsen notes that in the early twentieth century, professional advancement for dentists meant attempting to distance 'themselves, their organizations, journals and societies from the commercial aspects of their businesses'.[22] In 1930 Charles Johnson, the editor of the *Journal of the American Dental Association*, put this quite clearly: 'No man can become a member of a regularly organized professional association if he resorts to public advertising, and no man attempts today to become a member as long as he advertises.'[23] Unable to run ads in newspapers or distribute pamphlets, dentists could rely on their X-ray equipment as a professionally sanctioned kind of advertising, attesting to a dentist's skill and training, at least once the patient was in the door.

While Americans did not usually see ads for individual dentists (or dental X-rays), they were increasingly bombarded with advertising for dental products like toothpaste and antiseptic mouthwash. Just like the public school programmes, these ads encouraged Americans to become consumers of dental care. As historian Alyssa Picard argues, 'personal-care product manufacturers aggressively promoted the idea that having a pleasant mouth was critically important to one's social and psychological success'.[24] During the Depression, with so many adults thrown out of work, the social promise of healthy teeth was even more urgent, but also increasingly unattainable. Many potential customers wished for dental care but were unable to pay. This motivated some dentists to support New Deal-style plans for national dental insurance, but overall, dentists as a group were suspicious of any plan for socialized dental care. Most continued to feel that dental charity should only be available to the very poorest (and those deemed 'worthy').[25] It wasn't until the 1950s and 1960s that American workers increasingly benefitted from private dental insurance through their employers.[26] So while the growth in dental consumer products nurtured new desires for dental care, without insurance, those desires were most often unrequited.

This meant that in the 1930s, professional dental care was still a luxury for most people. A display of white, healthy teeth was a sign of wealth and privilege, and maintaining or improving the appearance of teeth represented a path to upward class mobility. But this was a path accessible only to a small percentage of potential customers. X-ray companies promised dentists that their equipment would signal wealth and high social class and attract the right kinds of patients, ones who were wealthy enough to pay for dental services. Not so subtly, X-ray salespeople were coached to praise dentists on 'that little touch of refinement in their office. Betting that their patients have noticed the change.' They promised dentists:

> 'By the introduction of X-ray work into your practice you will find that you will unearth a higher class of work which will enable you to eliminate undesirables from your practice.'

Without an X-ray, 'the better class in the community ... will avoid [your] office'.[27] The front cover of one Ritter pamphlet shows a patient prominently displayed wearing pearls (Figure 1.1). This 'better class' is described more explicitly in another Ritter publication

Figure 1.1 A 1933 advertising pamphlet for the Ritter X-Ray Machine circulated to dentists.
Better Radiography with the Ritter X-Ray Machine, Ritter X-Ray Company, 1933

that notes: 'generally speaking, American-born people seek intelligent dental service more readily than do others'.[28] The racist coding here is clear, and images of both model patients and model dentists show a pervasive expectation of whiteness throughout these X-ray ads. Picard argues that early-twentieth-century school dental hygiene programmes linked healthy teeth to good citizenship, targeting immigrant neighbourhoods as a way of inculcating white, Protestant values.[29] But again, nurturing these values did not make dental care necessarily affordable or even available for all communities. One estimate from 1934 showed one dentist for about 1,700 people in the United States, with a better ratio of one to one thousand in large cities. However, given that this same source expected that one dentist could successfully care for five hundred people a year, there was clearly a national dentist shortage.[30] And access was uneven. For Black Americans, dental care was particularly difficult to obtain. In 1930 one estimate showed only one dentist per seven thousand Black Americans, and that number had dropped to one per nine thousand by 1940.[31] There was a sharp decline in Black students enrolling in dental school over the same period.[32] This was likely due to the effects of the Depression, which had made dental school unaffordable and dental practice less profitable, causing an overall decline in the number of dentists per capita in the 1930s.[33]

Both the expected patient and the expected dentist in X-ray advertisements were white, and companies further aligned dentists with their anticipated patients by acknowledging dentists' own desire for financial security. Ritter promised dentists that their machine 'puts patients in a frame of mind where they expect and are willing to pay for expert dental service'.[34] Dentists would be able to charge patients a fee for the X-ray and then for any treatment needed after the examination, 'which the patient will usually demand or accept upon the dentist's suggestion'.[35] In the Ritter *Sales Approaches* manual, salespeople were instructed to make the financial benefits very clear to dentists. 'Contract for excessive restorative work in the mouth is always made easy through means of radiographic evidence.' Acknowledging that, 'Patients want to know why they are spending money … No better means has been demonstrated than an X-ray examination made in your own office.' With an X-ray examination, patients could more easily be convinced of the necessity of dental work and 'a dentist need not be timid in asking for a just fee'.[36] X-ray machines could instill patient trust in dentists while also serving as a vehicle of social advancement for dentists just as much as for patients. In one short story circulated by Ritter to dentists, a fictional Dr Colby is able to finally buy Mrs Colby a fur coat thanks to the successful installation of a Ritter X-ray machine.[37]

Science, progress, modernity

The financial promise of X-ray machines rested on their ability to make the right impression on patients. As X-ray company Watson & Sons emphasized, 'Dental education is leading the public to demand better service from the dentist [who should] aim at cultivating a reputation for modern methods and up-to-date service.'[38] Medical equipment companies assured dentists that one of the best ways to signal these modern methods was through the right kinds of visual cues in their dental office. An ad for S.S. White Operating Equipment argued: 'Your equipment speaks. What will you make it say about you? The environment into which you invite your patient, subtly tells whether or not you are successful and progressive.'[39] Similarly, advertising for General Electric X-ray equipment told dentists: 'Frequent use of the X-ray is one of the ways by which the public is learning to distinguish the progressive dentist',[40] and Ritter agreed that 'A dentist is accepted by his patients as being as modern as his surroundings indicate.'[41] Standard visual means of diagnosis were dismissed as old-fashioned: 'A mere mouth mirror, explorer, cotton plier examination of the oral cavity, is as antiquated as a horse and buggy.'[42]

X-ray equipment was not the only prop in this performance of modern, scientific dentistry. Grumsen has analysed ads for dental products in this period and found that by the 1920s, mentioning 'science' became a regular feature in toothpaste ads, with dentists shown wearing white coats, looking into microscopes or lighting Bunsen burners, signifying their alignment with scientific dentistry.[43] Dentists were told that the public regarded dentistry as an exacting science, and expected these visual displays of scientific equipment and dress. But, of course, X-ray companies did their best to emphasize the special power of X-ray equipment in particular to create the right visual impression. 'The Ritter X-ray machine conveys the correct psychological effect on the patient, creating an impression of scientific assurance, in a manner which suggests itself as a masterpiece of equipment, associated with skill and knowledge on the part of the dentist.'[44] While the process of taking an X-ray and the resulting X-ray image were clearly crucial to a performance of modern dentistry, it was the machine itself, 'a masterpiece of equipment', that was the star of the show, even when it was not turned on.

Medical machines have not always functioned as clear symbols of legitimate or scientific medicine. In the nineteenth century, most doctors using electricity as a therapeutic agent struggled to gain professional recognition.[45] Even X-ray equipment embodied multiple kinds of meanings in the late nineteenth and early twentieth centuries, associated with a new, more powerful medical sight, but also with magic, ghosts, and titillating access to the hidden body.[46] As many historians have noted, X-rays were medical images and entertainment at the same time. Yet by the 1930s, X-rays were more firmly established in hospitals, and a stronger association with science and with progress had solidified. This cultural authority of X-rays as a scientific object fuelled the widespread commodification of X-ray technology.[47] In addition to encounters with X-rays in medical and dental clinics, American consumers could use an X-ray machine to check the fit of new shoes in department stores, or visit an X-ray epilation salon to remove unwanted hair.[48]

But just as manufacturers expected the imagined patient to demand scientific dentistry, they also expected that these 'patients know little about the scientific principles of dentistry'.[49] As Nancy Tomes has noted, 'medicine exposes the disconnect between market logics of personal choice and the need to trust and defer to experts'.[50] Where the machine and its operation might be inscrutable to patients, the X-ray images themselves could be shown to patients as proof of dental skill and expertise. X-ray ads promised dentists that their patients would 'appreciate the accuracy of an X-ray diagnosis'. As 'with their own eyes they can see the conditions that need treatment, and it is only natural that they should have greater confidence in the dentist'.[51] A dentist recommending a course of treatment could now point to an X-ray, something visible and tangible to show to sceptical or reluctant patients.

There is an important tension here in the way X-ray images functioned both as seemingly objective proof of the state of tooth anatomy but also reassurance of the dentist's skill and expert diagnosis. Patients were expected to be able to see tooth decay 'with their own eye', but as we see in Figure 1.2, the dentist is explaining the X-ray to the patient by pointing to particular parts of the image. The image did not speak for itself. Expert interpretation was key.

This same relationship of viewer to X-ray image holds in other spheres in this period. In the very early twentieth century, X-rays were brought into court cases as evidence to support eyewitness testimony with the expectation that the X-ray simply revealed evidence of the body and would be straightforward for a jury to interpret. An X-ray showing a bone fracture, for instance, might corroborate

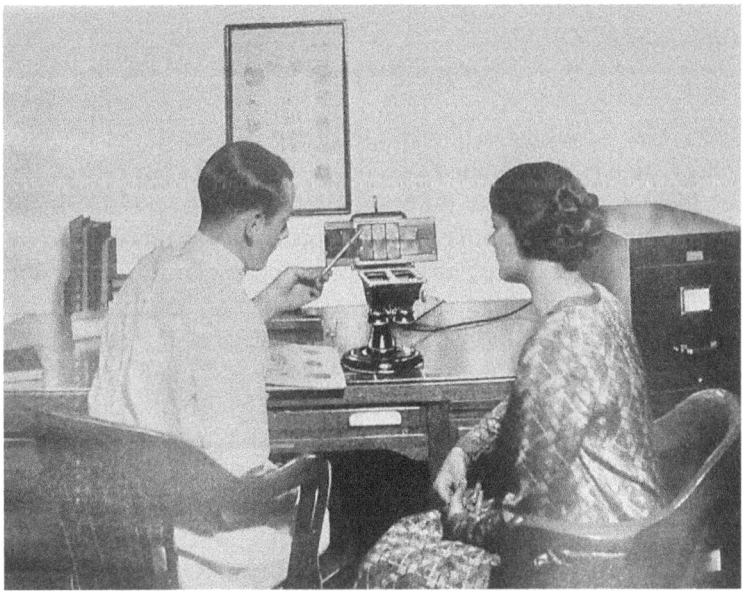

Figure 1.2 A patient being shown her dental X-rays in a Ritter
X-ray sales pamphlet.
Better Radiography with the Ritter X-Ray Machine, Ritter X-Ray
Company, 1933, p. 36

a plaintiff's claim that they had suffered an injury. Increasingly, however, expert witnesses were brought in to interpret the image for members of the jury and to attest to the process of creating the image.[52] Similarly, the practice of allowing patients to walk away with a copy of their medical X-rays faded fairly quickly in the early twentieth century. The first doctors to use X-rays insisted on their own interpretive expertise, wanting to distinguish their practice from the commercial and recreational X-rays available to members of the public who rushed to get their hands and feet radiographed in department stores and at fairs.[53]

Overall, then, the patient constructed in early X-ray advertisements was one who was discerning and demanding, able to pay for dental care, and likely to choose a dentist based on the impression of modernity, prosperity, science, and skill projected by both the X-ray equipment and its images. At the same time, this was a patient unable to fully evaluate the skill of the dentist, reliant on the expertise of the dentist to make sense of X-ray evidence.

Dental X-ray design

So far, I have considered how this imagined patient played a central role in companies' attempts to encourage dentists to buy their X-ray machines. But what about the design of the equipment? If the idea of this patient mattered to dentists choosing X-ray equipment, can we see any reflection of that ideal patient in particular features of the technology? While X-ray companies invoked an imagined patient who expected skill and expertise from their dentist, the design of actual X-ray equipment promised to take most decisions out of the hands of the dentist. Ritter very consciously pitched this disconnect to dentists in 1933, promising that their machine was: 'So simple and easy to operate that no special training is necessary, yet to the observer it suggests itself as a scientific appliance requiring the full knowledge and intelligence of the operator.'[54] Ritter promised that because of all the fixed and stable factors of operation, the resulting X-ray images were never too light or too dark, 'but always the proper density' (Figure 1.3).

One prominent feature of a number of dental X-ray machines was the automatic timer, which Ritter promised would 'eliminate

Never too Light   Never too Dark   But Always the Proper Density

FIXED FACTORS
that insure uniform detail in all films

Figure 1.3 Ritter promising that their new Model B machine delivers
consistently uniform X-rays every time.
An X-Ray Achievement by Ritter, Ritter X-Ray Company, 1933

all guesswork' for dentists. Dentists could rotate the dial on the
timer to the desired length of exposure, and the machine would
turn off automatically when the time was up. Figure 1.4 shows a
close-up of the timer on the Ritter machine, with a handy chart for
reference. These timers appear to have been standard features for
many dental X-ray manufacturers. On a similar timer for a machine
sold by Victor X-Ray Company, dentists could look up exposure
times depending on the part of the mouth – lower molars, upper
cuspids and bicuspids – the age of their patients – young person or
adult over or under fifty – and the kind of X-ray film, with options
for regular or extra-fast films. For each area of the mouth, a final
column also offered a recommended angle for the head of the X-ray
tube. Figure 1.5 shows a dial with this angle, and another useful
reference chart offering recommended angles for X-rays of different
kinds of teeth, and Figure 1.6 shows a dentist posing with one of
these timers.

Ritter promised that there was no need for dentists to hire
a specially trained graduate or specialist to help operate the
machine: 'There are no mysterious or cumbersome adjustments; in
fact everything is charted and all that is necessary is to follow a simple routine of exposure technique.'[55] Operators did not even need
to memorize these exposure times and angles, since the charts were
printed on the machine in multiple places. So, while these companies promised that patients would experience getting an X-ray as
proof of a dentist's special expertise, the design of the technology

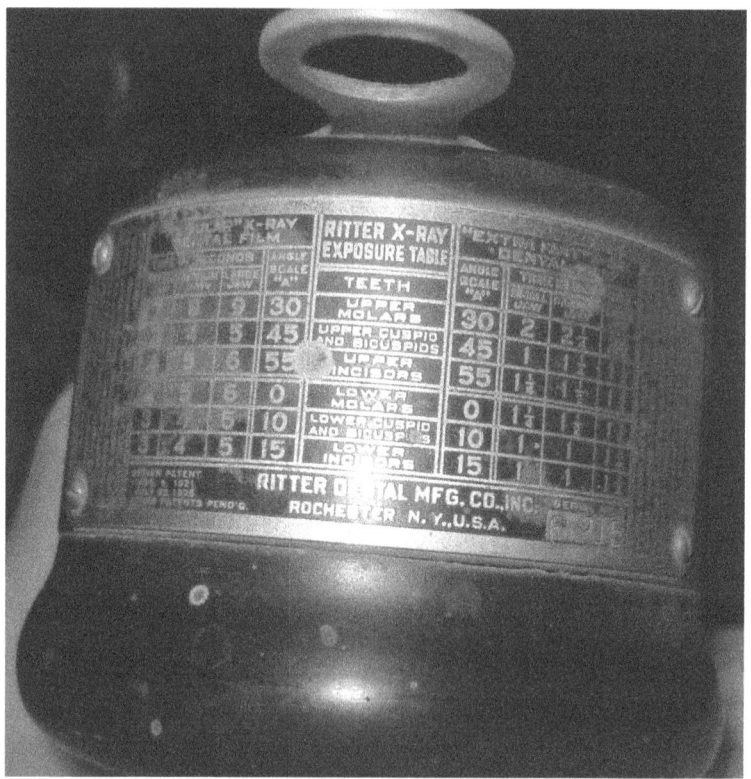

Figure 1.4 A reference chart on an automatic timer
made by Ritter Dental X-Ray Company.

itself emphasized that routine operation would require very little
judgement or skill.

Some publications even emphasized the possibility of delegating
almost all X-ray work to an assistant who could position the patient,
place the film in the patient's mouth, position the X-ray tube, make
the exposure, and develop the film. The dentist could then appear
at the end of this process to interpret the X-ray. This division of
labour was economically beneficial, allowing the dentist to spend
time on other billable procedures.[56] In the 1930s these dental workers were almost always women. Some were formally trained and
licenced dental hygienists, but the majority were assistants whose

Figure 1.5 An angle indicator on a Ritter machine with a chart showing
the correct angle for taking an X-ray of different teeth.

education and preparation varied. The first programmes for dental
hygienists were founded in the 1910s, and the American Dental
Hygienists Association was founded in 1923. By the early 1930s,
there were seventeen schools providing training for hygienists in the
United States, and surveys showed about 1,500 dental hygienists at
work in private practice along with almost fourteen thousand other
assistants and attendants.[57] A book published by the Women's
College of the University of North Carolina intending to introduce
women to different professional opportunities in dentistry noted
that both hygienists and unlicenced assistants should learn how to
use an X-ray machine and develop films.[58] Both Eastman Kodak
and Ritter offered classes so that dental assistants could learn X-ray
techniques.[59]

While dentists might delegate the operation of the machine,
they were responsible for setting up the equipment, a process that
required careful attention to detail and some expertise with electrical
equipment. The installation and user manual for the Ritter machine
looked very different from the advertising pamphlets. Pictures of
gleaming, streamlined equipment were replaced with schematic

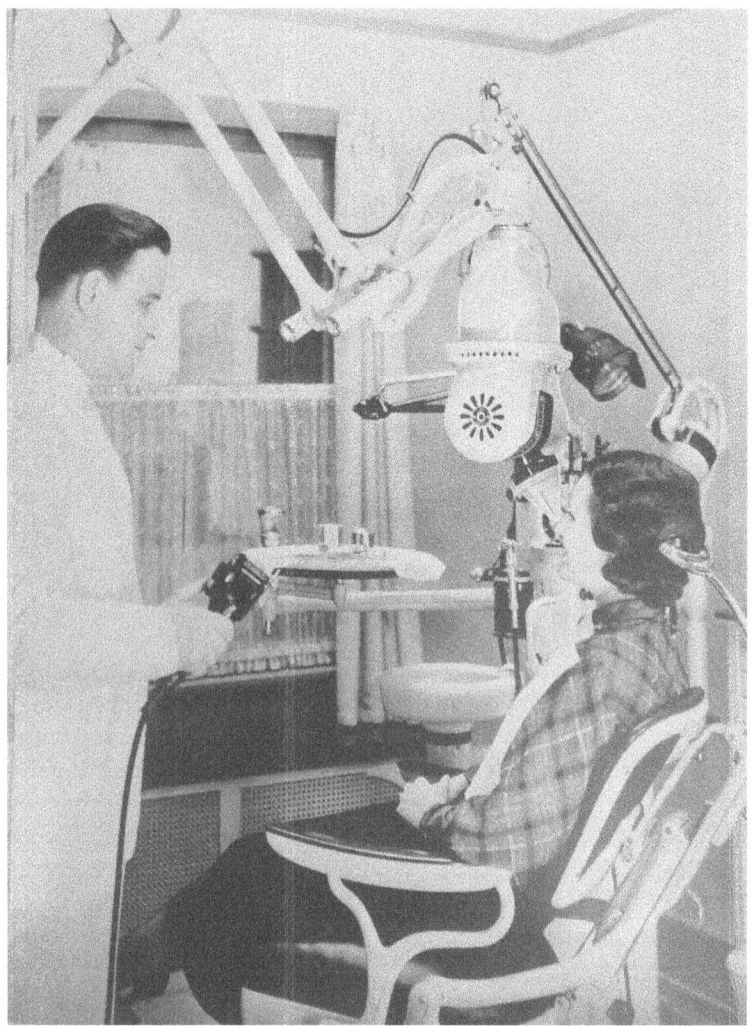

Figure 1.6 A dentist posed in a Ritter sales pamphlet with
an automatic timer in his hand.
An X-Ray Achievement by Ritter, Ritter X-Ray Company, 1933

drawings covered in labelled parts (see Figure 1.7). Upon receiving
a machine, dentists were faced with many pages of instructions,
and were admonished to read these carefully before assembling and
operating the machines. Dentists were warned, for instance, that

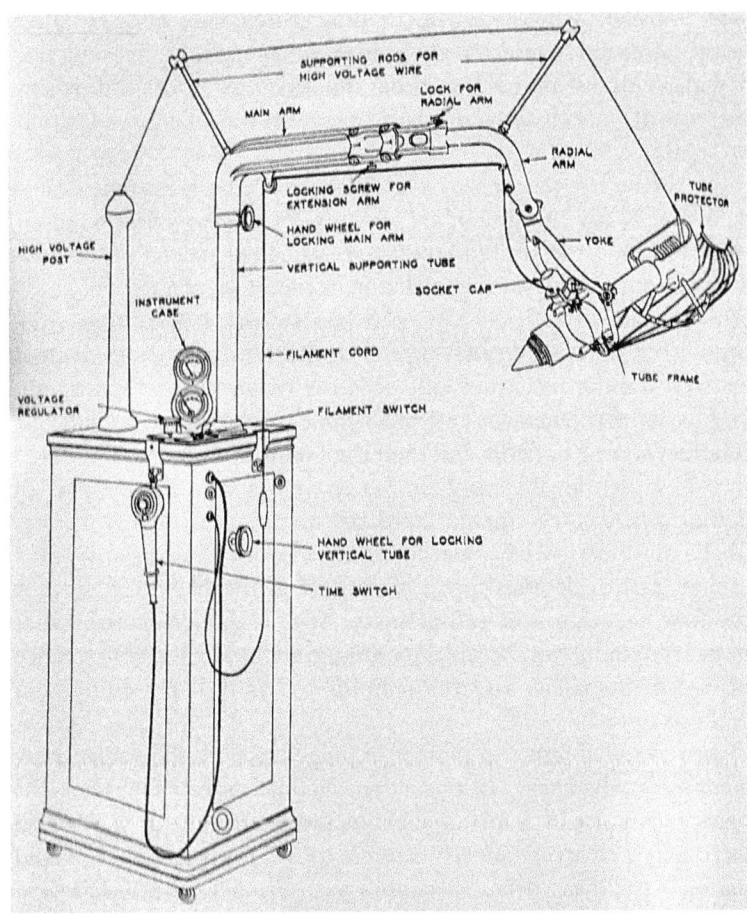

Figure 1.7 A schematic of a Ritter dental X-ray machine.
Operator's Manual for the Ritter X-Ray Machine, Rochester,
NY: Ritter Dental Manufacturing Co., 1933, p. 6

the milliampere meter always read from one to two milliamperes higher when the X-ray tube was cold, and so they needed to wait to allow the tube to warm up to produce the desired intensity of X-rays.[60] Even when the voltage and current were correct, dentists needed to do more than simply read an angle and length of exposure off of a chart. This was complex electrical equipment, and dentists needed to watch out for possible electrocution. They were warned: 'when pressing the time switch, care must be taken not to come into contact with the high voltage wire'.[61] Tapping into a wider cultural desire for speed and efficiency,[62] the advertising promise of smooth and automatic operation was of course not fully attainable.

Nevertheless, the step-by-step instructions built into the design of dental X-ray equipment are striking. By the early 1920s, many universities in the United States had launched special postgraduate training for doctors specializing in radiology, courses of study which included at least a year of coursework before apprenticing with a practising radiologist.[63] In hospital medicine, doctors resisted standardized methods of X-ray diagnosis and treatment, emphasizing the necessity of their clinical art and judgement due to idiosyncratic patients and their own unique X-ray equipment.[64] In a busy dental office, where X-ray work was only one part of dental practice, a more standardized approach to this work was likely quite attractive. Special knowledge of X-rays was not a crucial part of dentists' professional identities the way it was for the first generation of radiologists. Add in the expectation that patients would not be able to judge much of the performance of X-ray diagnosis, and the simplified dental X-ray equipment makes sense.

Just as the ability of a patient to judge the skill of a dentist only went so far, the agency of this patient had limits as well. There was agency in choice of dentist and clinic, but once in the door, patients were expected to obediently submit to an X-ray exam. Ads and manuals for these dental machines are full of images of patients, most often women, patiently sitting motionless for their X-rays, holding their heads and bodies still and in just the right position (Figure 1.8). While Ritter boasted of all the factors of operation of the machine that they had standardized – timing, voltage, angles – they barely mentioned the need to discipline the patient to maintain a particular kind of posture. A 1933 instruction book leads dentists through careful steps for positioning patients, film, and the cone of the X-ray machine depending on which teeth are being radiographed. At one point, dentists are told that 'the patient should be cautioned not to tilt the head to either side', but overall concerns about fidgeting or uncooperative patients are not present.[65] The compliance of the patient is simply expected.

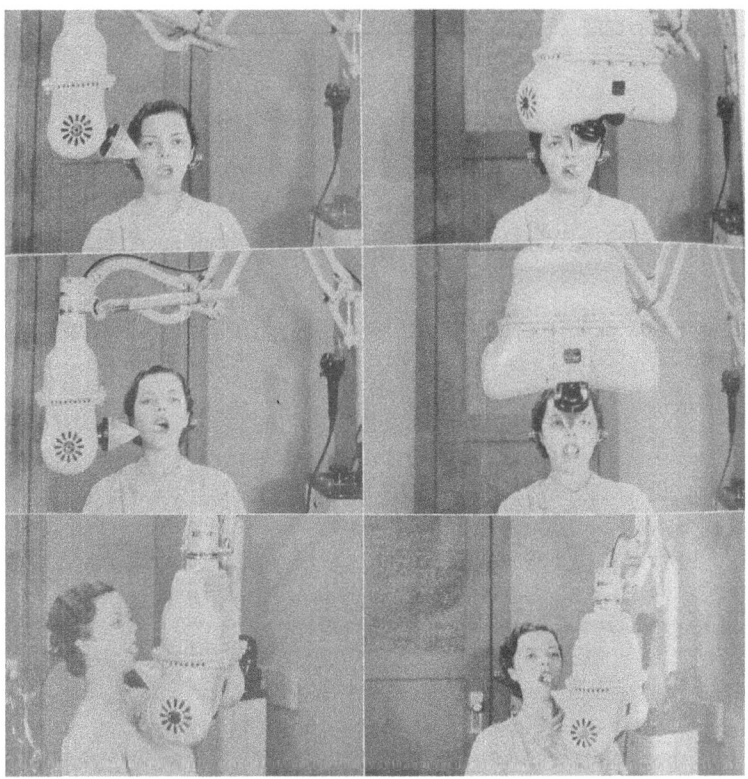

Figure 1.8 A model patient sitting for dental X-rays.
An X-Ray Achievement by Ritter, Ritter X-Ray Company, 1933

Safety and trust in the machine

The extent of this necessary trust on the part of the patient is highlighted even more when we consider issues of safety. Walking into
any room with an X-ray machine was immediately risky. X-ray
equipment could cause electrical shocks, patients and dentists could
be hit by falling or unstable equipment, and overexposure to X-ray
radiation could cause burns, hair loss, blood changes, and, eventually, cancers. Even worse, there was usually no immediate sign of
radiation exposure. Symptoms were often delayed and could take
weeks or even years to develop. These X-ray dangers were well
known to doctors and to the wider public by the 1930s, though still

in a framework that assumed that there was a minimum safe level of radiation exposure.[66]

X-ray marketing attempted to pre-empt some of these anxieties. Ritter salespeople were coached to anticipate dentists asking: 'What is all this danger propaganda that I read about in connection with X-ray apparatus?' The appropriate reply: 'Doctor, there is absolutely no danger in the use of a Ritter Shock-Proof X-ray unit', which is 'electrically, mechanically and radiologically safe'.[67] Ritter salespeople needed to assuage dentists' fears about harming their patients as well as fears about their own safety. To minimize concern about patient safety, salespeople were told to mention that over one hundred machines were already in use, making fifteen thousand to sixty thousand exposures per year, and 'There has never been a report of an accident from electric shock or X-ray exposure with a Ritter Model "B" X-ray unit.' Besides denying all possibility of harm, another tactic was to point out to dentists the non-radiological dangers of dentistry, including fractured jaws from surgical extractions and patients inhaling plaster into their lungs when tooth impressions were taken.[68]

Even without any kind of accident or error operating the machines, patients were guaranteed to receive a direct dose of X-rays. Dentists were assured of their own safety since 'actual exposures are really confined to the patient only, who becomes the absorbing medium'.[69] By this time, doctors had developed multiple kinds of measurement techniques for measuring the dose of X-rays reaching a patient. Dentists, however, were asked to simply trust the manufacturers that the dose delivered to their patients was within safe limits. In the Ritter *Text Book of X-Ray Technique*, there is one very short discussion of dosage and dentists are told not to exceed half of an erythema dose, which is half of a dose necessary to cause reddening of the skin of the patient. Ritter assures readers: 'the exposures we have recommended are well within the limit'.[70]

Dentists were asked to place the safety of their patients' bodies in Ritter's hands. They were also asked to trust that their own bodies would be safe. There is no mention of radiation safety in a 1929 Ritter X-ray user manual.[71] The 1933 *Text Book of X-Ray Technique* published by Ritter does emphasize under 'Important Notes' that the 'operator should not stand in the direct path of rays when making exposures', and in fact should stand as far away as the cord on the timer will permit.[72] This is despite the fact that Ritter promised that 'complete protection' had been achieved by enclosing the X-ray tube in a lead glass shield of 1/16th inch thickness. No explanation was given as to why that thickness was sufficient, but it did conform to recommendations published in a handbook on X-ray safety produced by the National Bureau of Standards (NBS) in 1931. Looking up the recommended voltage of the tube (75–100 volts) alongside the 1931 NBS guidelines, and converting inches to millimetres of lead, I can confirm that the level of shielding does match those guidelines.[73] However, it is unclear whether dentists had those guidelines at their fingertips, and the NBS handbook was not mentioned in any Ritter manual from the early 1930s. In the conversations leading up to the publication of these NBS guidelines, there was an initial proposal to have one representative from a dental association on the American committee responsible for drafting these safety protocols.[74] However, in the end the assembled committee did not include a dentist, and the NBS guidelines were clearly intended for hospital X-ray rooms. The guidelines mention dental offices once, and only to exclude them from certain protocols: 'All X-ray rooms (except for dental radiography) shall be lined throughout with sheet lead.'[75] It's not clear why dental X-ray set-ups were exempt from this recommendation, though it suggests agreement with X-ray manufacturers that radiation produced by dental equipment was within safe limits.

The invisibility of dental spaces in these guidelines meant that dentists took on a much less active role in ensuring X-ray safety compared to doctors. X-ray safety in hospitals remained the responsibility of doctors, who monitored lead shielding for cracks and supervised routine blood counts for their technicians. This was in part an acknowledgement that the quantitative rules about necessary thicknesses of lead shielding masked deep uncertainties about the actual effects of radiation on bodies. The numbers were based on a best guess of what was safe, extrapolated from observations of X-ray rooms in British hospitals in the 1920s.[76] However, despite these uncertainties, there is no sense that dentists needed to be actively double-checking for safety. Ritter promised that responsibility for safety rested in the machine itself.

So what happened to our imagined patient? At first this was an active, discerning, and even demanding consumer, one who chose a dentist based on an impression of skill and expertise. This patient was the crucial audience for the dentist-as-expert, performing difficult, scientific dentistry. But this same patient, once in the dentist's chair, became passive and trusting, submitting quietly to an X-ray examination, and holding their head in just the right way to be radiographed, without concern for their safety. Perhaps it was the image of this transformed and docile patient that gave dentists permission to abdicate most responsibility and judgement to the equipment itself. Patients were expected to place their trust in the dentist, and dentists were expected to place their trust in the machine.

Conclusion

Coming back to the initial questions posed by this chapter about the imagined patient and the design of this technology, I've found that dental X-ray equipment in the 1930s emphasized easy, safe, and routine use for dentists, and expected passive, compliant patients. Yet X-ray equipment was also expected to be a crucial prop in the performance of scientific dentistry, advertising dental skill and expertise for an imagined consumer who demanded modern, progressive dentistry. This is not to downplay the actual benefits of the kinds of diagnoses made possible with X-rays – dentists were delighted to be able to see cavities forming months before they were evident in a visual examination. But at a time when advertising directly to patients was professionally taboo for dentists, we need to consider the important advertising work done by the machines themselves, and the disjunction between what they promised to patients and the actual experience of operating one of them.

It is hard to know whether new X-ray equipment in the 1930s and 1940s did indeed draw in new dental customers, or make X-ray consumers out of existing patients. This was still a period before employer-sponsored dental insurance made dental care affordable for a greater percentage of American consumers. What is clearer is that this was the period in which X-ray came to be seen by dentists as an indispensable clinical technology. By 1940 the Dental Educational Council of the American Dental Association affirmed that they expected to see roentgenology included as a required part of the dental school curriculum.[77] And by the mid-1940s, X-ray equipment was no longer seen as a discretionary or luxury purchase for a new dentist. Dentists setting up a new practice were told: 'The X-ray machine is an extremely important item and should be used unsparingly. If necessary, the beginner should economize on the chair, cabinet and other accessories in order to purchase good X-ray equipment.'[78] As this technology became ubiquitous in dental offices, the power of the particular impression it could create for patients likely faded a bit as well. The X-ray remained a symbol of scientific dentistry, but as a baseline expectation of competence rather than a mark of special expertise.

Notes

1 Joel D. Howell, *Technology in the Hospital: Transforming Patient Care in the Early Twentieth Century* (Baltimore, MD: Johns Hopkins University Press, 1995); Ullmann, H.J., 'Radiation Dosage: Standardization versus Individual Adaptation.' *Radiology* 1 (1923): 31–3.

2 Milton Asbell, *Dentistry: A Historical Perspective* (Bryn Mawr, PA: Dorrance & Company, 1988), pp. 140–1.

3 Joseph A. Pollia, 'The Alveolar Radiogram: What It Tells.' Journal of the American Dental Association 9 (1922): 36–51, 38; James David McCoy, *Dental and Oral Radiography: A Textbook for Students and Practitioners of Dentistry*, 3rd edition (St. Louis: C.V. Mosby Co., 1922), G.V. Black, *A Work on Operative Dentistry: The Technical Procedures in Filling Teeth*, 6th edition, Vol. 2 (Chicago, IL: Medico-Dental Publishing Co., 1924), p. 376.

4 Victor X-Ray, 'Advertisement.' *Journal of the American Dental Association* 9, no. 12 (1922): A-7.

5 Ritter X-Ray Company, *Sales Approaches* (Rochester, NY: Ritter X-Ray Company, 193–).

6 Woodhouse, Chase Going, and Ruth Yeomans Schiffman. Dentistry as a Profession (Greensboro, NC: Women's College of the University of North Carolina, 1934), pp. 14–15.

7 Chase Going Woodhouse and Ruth Yeomans Schiffman, *Dentistry: Its Professional Opportunities* (Greensboro, NC: Women's College of the University of North Carolina, 1934), p. 58.

8 A.B. William Suter, *Dentistry: A Profession and a Business* (Rochester, NY: Ritter Dental Manufacturing Co., 1930), pp. 71, 93, 112.

9 Ritter X-Ray Company, *Sales Approaches*.

10 Ritter X-Ray Company, *Sales Approaches.*

11 Ruth Schwartz Cowan, 'The Consumption Junction: A Proposal for Research Strategies in the Sociology of Technology', in *The Social Construction of Technological Systems*, edited by Wiebe E. Bijker, Thomas P. Hughes, and Trevor Pinch, pp. 261–80 (Cambridge, MA: MIT Press, 1987), p. 273.

12 Trevor Pinch and Wiebe E. Bijker, 'The Social Construction of Facts and Artifacts', in Bijker, Hughes, and Pinch, *The Social Construction of Technological Systems*, pp. 17–50.

13 Ritter was one of the major manufacturers and suppliers of dental equipment, including dental X-ray apparatus, in this period. The National Museum of American History archives have a particularly rich set of trade literature from Ritter including advice for salespeople, multiple kinds of advertising pamphlets for dentists, and longer manuals and textbooks explaining X-ray operation. While I haven't located actual sales data, this archival material suggests a widespread advertising campaign that likely reached a large audience of dentists.

14 Alyssa Picard, *Making the American Mouth: Dentists and Public Health in the Twentieth Century* (Piscataway, NJ: Rutgers University Press, 2009), p. 22.

15 Evangeline M. Jordan, *Operative Dentistry for Children* (Brooklyn, NY: Dental Items of Interest Publishing Co., 1925), p. 31.

16 United States Senate, Committee on Education and Labor. *Dental Research and Dental Care: Hearings before a Subcommittee of the Committee on Education and Labor* (Washington, DC: United States Government Printing Office, 1945), p. 88.

17 Picard, *Making the American Mouth*, pp. 18–27.

18 Quoted in Picard, *Making the American Mouth*, p. 23.

19 Committee on the Future of Dental Education, 'Evolution of Dental Education', in *Dental Education at the Crossroads: Challenges and Change*, edited by Marilyn J. Field, pp. 35–58 (Washington, DC: National Academy Press, 1995).

20 S.S. White Operating Equipment. 'Advertisement.' *The Dental Review* (1932): 137.

21 Watson & Sons, *Radiography for the Dentist* (Kingsway, London: Watson & Sons, 192–), p. 6.

22 Stine Grumsen, 'Zeal of Acceptance: Balancing Image and Business in Early Twentieth-century American Dentistry.' *Medicine Studies* 3 (2012): 197–214, 199.

23 Quoted in Grumsen, 'Zeal of Acceptance', 200.

24 Picard, *Making the American Mouth*, p. 106.

25 Picard, *Making the American Mouth*, Chapter 4.

26 Picard, *Making the American Mouth*, Chapter 6.

27 Ritter X-Ray Company, *Sales Approaches*.

28 Suter, *Dentistry*, p. 37.

29 Picard, *Making the American Mouth*. Tracey Adams has uncovered a similar dynamic in Ontario. See Tracey Adams, *A Dentist and a Gentleman: Gender and the Rise of Dentistry in Ontario* (Toronto: University of Toronto Press, 2000).

30 Woodhouse and Yeomans Schiffman, *Dentistry as a Profession*, pp. 3–6.

31 Picard, *Making the American Mouth*, p. 113.

32 Clifton O. Dummett, 'The Negro in Dental Education.' *The Phylon Quarterly* 20, no. 4 (1959): 379–88, 381.

33 The 1930 census showed one dentist per 1,729 people, and in 1940 that had dropped to one dentist per 1,865 people: *Dental Research and Dental Care*, p. 19.

34 Ritter X-Ray Company, *Better Radiography with the Ritter X-Ray Machine* (Rochester, NY: Ritter X-Ray Company, 1933), p. 30.

35 Ritter X-Ray Company, *Sales Approaches*.

36 Ritter X-Ray Company, *Sales Approaches*.

37 Ritter X-Ray Company, *Mrs. Colby's Fur Coat* (Rochester, NY: Ritter X-Ray Company, 1935).

38 Watson & Sons, *Radiography for the Dentist*, 5.

39 S.S. White Operating Equipment, 'Advertisement', 137.

40 General Electric, 'Advertisement.' *The Dental Review* (1931): 99.

41 Ritter X-Ray Company, 'Advertisement.' *The Dental Review* (1931): 96.

42 Suter, *Dentistry*, p. 325.

43 Stine Grumsen, 'The Era of Whiter Teeth: Advertising in American Dentistry 1910–1950.' *Journal of the History of Dentistry* 57, no. 2 (2009): 75–84, 77.

44 Ritter X-Ray Company, *Better Radiography with the Ritter X-Ray Machine*, p. 3.

45 Vivien Hamilton, 'Medical Machines as Symbols of Science? Promoting Electrotherapy in Victorian Canada.' *Technology and Culture* 58, no. 4 (2017): 1017–45; Iwan Rhys Morus, 'Batteries, Bodies and Belts: Making Careers in Victorian Medical Electricity', in *Electric Bodies: Episodes in the History of Medical Electricity*, edited by Paola Bertucci and Giuliano Pancaldi, pp. 209–38 (Bologna: Universita di Bologna, 2001).

46 Lisa Cartwright, *Screening the Body: Tracing Medicine's Visual Culture* (Minneapolis, MN: University of Minnesota Press, 1995).

47 Matthew Lavine, *The First Atomic Age: Scientists, Radiations, and the American Public,* 1895–1945 (Basingstoke: Palgrave Macmillan, 2013).

48 Jacalyn Duffin and Charles Hayter, 'Baring the Sole: The Rise and Fall of the Shoe-Fitting Fluoroscope.' *Isis* 91 (2000): 260–82; Rebecca Herzig, 'Removing Roots: "North American Hiroshima Maidens" and the X-ray.' *Technology and Culture* 40, no. 4 (1999): 723–45.

49 Ritter X-Ray Company, *Better Radiography with the Ritter X-Ray Machine*, p. 36.

50 Nancy Tomes, *Remaking the American Patient: How Madison Avenue and Modern Medicine Turned Patients into Consumers* (Chapel Hill, NC: University of North Carolina Press, 2016), p. 10.

51 Ritter X-Ray Company, *Better Radiography with the Ritter X-Ray Machine*, p. 36.

52 Tal Golan, 'The Emergence of the Silent Witness: The Legal and Medical Reception of X-rays in the USA.' *Social Studies of Science* 34, no. 4 (2004): 469–99.

53 Howell, *Technology in the Hospital*, p. 197.

54 Ritter X-Ray Company, *Better Radiography with the Ritter X-Ray Machine*, p. 4.

55 Ritter X-Ray Company, *Sales Approaches.*

56 Suter, *Dentistry*, pp. 292–3.

57 Woodhouse and Schiffman, *Dentistry*, pp. 105–6.

58 Woodhouse and Schiffman, *Dentistry*, p. 115. For a discussion of gendered division of dental labour in one Canadian context, see Adams, *A Dentist and a Gentleman*, pp. 110–25.

59 Ritter X-Ray Company, *The Dentist's Diplomat* (Rochester, NY: Ritter X-Ray Company, 1930), p. 44.

60 Ritter X-Ray Company, *Operator's Manual for the Ritter X-Ray Machine* (Rochester, NY: Ritter X-Ray Company, 1929), p. 20.

61 Ritter X-Ray Company, *Operator's Manual for the Ritter X-Ray Machine*, p. 11.

62 Howell, *Technology in the Hospital*, pp. 30–68.

63 'Graduate Instruction in Radiology.' *American Journal of Roentgenology* 9 (1922): 465–7; Louis B. Wilson, 'Graduate Education in Roentgenology.' *American Journal of Roentgenology* 7 (1920): 580–1.

64 Ullmann, 'Radiation Dosage.'

65 Ritter X-Ray Company, *Instruction Book: Embodying the Use of the Modern Dental X-Ray Unit* (Rochester, NY: Ritter X-Ray Company, 1933), p. 20.

66 The idea that there was a minimum safe dose of radiation was increasingly contested following the nuclear attacks on Japan in World War

II, as well as research in the 1950s showing increases in paediatric leukaemia due to prenatal X-rays: Lavine, *The First Atomic Age*; J. Samuel Walker, *Permissible Dose: A History of Radiation Protection in the Twentieth Century* (Berkeley, CA: University of California Press, 2000); Gayle Greene, *The Woman Who Knew Too Much: Alice Stewart and the Secrets of Radiation* (Ann Arbor, MI: University of Michigan Press, 1999).

67 Ritter X-Ray Company, *Sales Approaches*.
68 Ritter X-Ray Company, *Sales Approaches*.
69 Ritter X-Ray Company, *Sales Approaches*.
70 Ritter X-Ray Company, *A Text Book of X-Ray Technique* (Rochester, NY: Ritter X-Ray Company, 1933), p. 60.
71 Ritter X-Ray Company, *Operator's Manual for the Ritter X-Ray Machine*.
72 Ritter X-Ray Company, *A Text Book of X-Ray Technique*, p. 28.
73 National Bureau of Standards, *X-Ray Protection*, Handbook No. 15 (Washington: United States Government Printing Office, 1931).
74 Taylor to Dr E.H. Skinner (29 October 1926, Lauriston Sale Taylor Archives, Box 8, Folder 21).
75 National Bureau of Standards, *X-Ray Protection*, p. 3.
76 Vivien Hamilton, 'X-Ray Protection in American Hospitals', in *Inevitably Toxic: Historical Perspectives on Contamination, Exposure, and Expertise*, edited by Brinda Sarathy, Vivien Hamilton, and Janet Farrell Brodie, pp. 23–49 (Pittsburgh, PA: University of Pittsburgh Press, 2018)
77 Harlan Hoyt Horner, *Dental Education Today* (Chicago, IL: University of Chicago Press, 1947), p. 18.
78 William McGehee and Alfred Walker, *Dental Practice Management* (Chicago, IL: The Year Book Publisher Inc., 1944), p. 55.

Bibliography

Adams, Tracey. *A Dentist and a Gentleman: Gender and the Rise of Dentistry in Ontario*. Toronto: University of Toronto Press, 2000.
Asbell, Milton. *Dentistry: A Historical Perspective*. Bryn Mawr, PA: Dorrance & Company, 1988.
Bijker, Wiebe E., and Trevor Pinch. 'The Social Construction of Facts and Artifacts', in *The Social Construction of Technological Systems*, edited by Wiebe E. Bijker, Thomas Hughes, and Trevor Pinch, pp. 17–50. Cambridge, MA: MIT Press, 1987.
Black, G.V. *A Work on Operative Dentistry: The Technical Procedures in Filling Teeth*, 6th edition, Vol. 2. Chicago, IL: Medico-Dental Publishing Co., 1924.

Cartwright, Lisa. *Screening the Body: Tracing Medicine's Visual Culture.* Minneapolis, MN: University of Minnesota Press, 1995.

Committee on the Future of Dental Education. 'Evolution of Dental Education', in *Dental Education at the Crossroads: Challenges and Change*, edited by Marilyn J. Field, pp. 35–58. Washington, DC: National Academy Press, 1995.

Duffin, Jacalyn, and Charles Hayter. 'Baring the Sole: The Rise and Fall of the Shoe-Fitting Fluoroscope.' *Isis* 91 (2000): 260–82.

Dummett, Clifton O. 'The Negro in Dental Education.' *The Phylon Quarterly* 20, no. 4 (1959): 379–88.

General Electric. 'Advertisement.' *The Dental Review* (1931): 99.

Golan, Tal. 'The Emergence of the Silent Witness: The Legal and Medical Reception of X-Rays in the USA.' *Social Studies of Science* 34, no. 4 (2004): 469–99.

'Graduate Instruction in Radiology.' *American Journal of Roentgenology* 9 (1922): 465–7.

Greene, Gayle. *The Woman Who Knew Too Much: Alice Stewart and the Secrets of Radiation.* Ann Arbor, MI: University of Michigan Press, 1999.

Grumsen, Stine. 'The Era of Whiter Teeth: Advertising in American Dentistry 1910–1950.' *Journal of the History of Dentistry* 57, no. 2 (2009): 75–84.

Grumsen, Stine. 'Zeal of Acceptance: Balancing Image and Business in Early Twentieth-Century American Dentistry.' *Medicine Studies* 3 (2012): 197–214.

Hamilton, Vivien. 'Medical Machines as Symbols of Science? Promoting Electrotherapy in Victorian Canada.' *Technology and Culture* 58, no. 4 (2017): 1017–45.

Hamilton, Vivien. 'X-Ray Protection in American Hospitals', in *Inevitably Toxic: Historical Perspectives on Contamination, Exposure, and Expertise*, edited by Brinda Sarathy, Vivien Hamilton, and Janet Farrell Brodie, pp. 23–49. Pittsburgh, PA: University of Pittsburgh Press, 2018.

Herzig, Rebecca. 'Removing Roots: "North American Hiroshima Maidens" and the X-Ray.' *Technology and Culture* 40, no. 4 (1999): 723–45.

Horner, Harlan Hoyt. *Dental Education Today.* Chicago, IL: University of Chicago Press, 1947.

Howell, Joel D. *Technology in the Hospital: Transforming Patient Care in the Early Twentieth Century.* Baltimore, MD: Johns Hopkins University Press, 1995.

Jordan, Evangeline M. *Operative Dentistry for Children.* Brooklyn, NY: Dental Items of Interest Publishing Co., 1925.

Lavine, Matthew. *The First Atomic Age: Scientists, Radiations, and the American Public, 1895–1945.* Basingstoke: Palgrave Macmillan, 2013.

McCoy, James David. *Dental and Oral Radiography: A Textbook for Students and Practitioners of Dentistry*, 3rd edition. St. Louis, MO: C.V. Mosby Co., 1922.

McGehee, William, and Alfred Walker. *Dental Practice Management*. Chicago, IL: Year Book Publisher Inc., 1944.

Morus, Iwan Rhys. 'Batteries, Bodies and Belts: Making Careers in Victorian Medical Electricity', in *Electric Bodies: Episodes in the History of Medical Electricity*, edited by Paola Bertucci and Giuliano Pancaldi, pp. 209–38. Bologna: Universita di Bologna, 2001.

National Bureau of Standards. *X-Ray Protection*. Washington: United States Government Printing Office, 1931.

Picard, Alyssa. *Making the American Mouth: Dentists and Public Health in the Twentieth Century*. Piscataway, NJ: Rutgers University Press, 2009.

Pollia, Joseph A. 'The Alveolar Radiogram: What It Tells.' *Journal of the American Dental Association* 9 (1922): 36–51.

Ritter X-Ray Company. 'Advertisement.' *The Dental Review* (1931): 96.

Ritter X-Ray Company. *Better Radiography with the Ritter X-Ray Machine*. Rochester, NY: Ritter X-Ray Company, 1933.

Ritter X-Ray Company. *The Dentist's Diplomat*. Rochester, NY: Ritter X-Ray Company, 1930.

Ritter X-Ray Company. *Instruction Book: Embodying the Use of the Modern Dental X-Ray Unit*. Rochester, NY: Ritter X-Ray Company, 1933.

Ritter X-Ray Company. *Mrs. Colby's Fur Coat*. Rochester, NY: Ritter X-Ray Company, 1935.

Ritter X-Ray Company. *Operator's Manual for the Ritter X-Ray Machine*. Rochester, NY: Ritter X-Ray Company, 1929.

Ritter X-Ray Company. *A Text Book of X-Ray Technique*. Rochester, NY: Ritter X-Ray Company, 1933.

Ritter X-Ray Company. *Sales Approaches*. Rochester, NY: Ritter X-Ray Company, 193–.

Schwartz Cowan, Ruth. 'The Consumption Junction: A Proposal for Research Strategies in the Sociology of Technology', in *The Social Construction of Technological Systems*, edited by Wiebe E. Bijker, Thomas P. Hughes, and Trevor Pinch, pp. 261–80. Cambridge, MA: MIT Press, 1987.

S.S. White Operating Equipment. 'Advertisement.' *The Dental Review* (1932): 137.

Suter, A.B. William. *Dentistry: A Profession and a Business*. Rochester, NY: Ritter X-Ray Company, 1930.

Tomes, Nancy. *Remaking the American Patient: How Madison Avenue and Modern Medicine Turned Patients into Consumers*. Chapel Hill, NC: University of North Carolina Press, 2016.

Ullmann, H.J. 'Radiation Dosage: Standardization versus Individual Adaptation.' *Radiology* 1 (1923): 31–3.

United States Senate, Committee on Education and Labor. *Dental Research and Dental Care: Hearings before a Subcommittee of the Committee on Education and Labor*. Washington, DC: United States Government Printing Office, 1945.

Victor X-Ray. 'Advertisement.' *Journal of the American Dental Association* 9, no. 12 (1922): A-7.

Walker, J. Samuel. *Permissible Dose: A History of Radiation Protection in the Twentieth Century*. Berkeley, CA: University of California Press, 2000.

Watson & Sons. *Radiography for the Dentist*. Kingsway, London: Watson & Sons, 192–.

Wilson, Louis B. 'Graduate Education in Roentgenology.' *American Journal of Roentgenology* 7 (1920): 580–1.

Woodhouse, Chase Going, and Ruth Yeomans Schiffman. *Dentistry: Its Professional Opportunities*. Greensboro, NC: Women's College of the University of North Carolina, 1934.

Woodhouse, Chase Going, and Ruth Yeomans Schiffman. *Dentistry as a Profession*. Greensboro, NC: Women's College of the University of North Carolina, 1934.

2

Chronic neglect: Race, dialysis, and vulnerable patienthood

Richard M. Mizelle, Jr

Dialysis patients in our society are both visible and invisible. Inconspicuous and unremarkable in their disabilities, they disappear three times per week for several hours, connected to a medical technology that is required for life. Over 500,000 people in the United States and two–three million worldwide receive dialysis.[1] Dialysis performs the work of kidneys by removing waste products from blood outside of the body and returning filtered blood to the body, prolonging the lives of people whose kidneys are no longer functioning properly. To enable enough blood to be filtered by the dialysis machine, vascular surgeons create an arteriovenous (AV) fistula or shunt by connecting an artery to a vein in the patient's arm. This forms a wider blood vessel and provides an easier access point for the two needles necessary for dialysis. The first inserted needle removes blood and sends it through the dialyser, and the second needle returns the cleansed blood back into the body. The dialyser is encased in a tube roughly a foot long and two to three inches wide with openings on both sides of the structure.[2] In a process called diffusion, blood enters the tube and is separated from dialysate by a thin semipermeable membrane that serves as a filter. Dialysate is a liquid solution composed of water, acid, bicarbonates, and electrolytes. Waste products in the blood (urea and creatinine) are pulled into the solution while the necessary blood cells and protein remain in the blood.[3] A tea bag in a hot cup of water is helpful for visualizing this diffusion process. The bag itself becomes the semipermeable membrane that allows the colour and flavour to pass through into the water but not the larger tea leaves.

The United States has the highest number of kidney failures per capita in the world, due in large part to rising rates of hypertension, diabetes, obesity, and other chronic diseases and conditions. Blacks and Latinos, moreover, make up the majority of those diagnosed with chronic kidney disease (CKD) and end-stage renal disease (ESRD), and rates of dialysis are only expected to increase as a large percentage of the population continues to age.[4] Dialysis is, in fact, both a life-saving technology and a chronic and perpetual life support – a slow-moving emergency that occurs three times per week for the remainder of most dialysis patients' lives. It is also an evolving technology with a complex history.

This chapter begins by contextualizing the history of dialysis in the United States in the 1960s and the inclusion of dialysis treatment under Medicare in 1972. From the inception of dialysis in the 1940s to the establishment of the Seattle Artificial Kidney Center (SAKC) in 1961 as the first outpatient dialysis centre in the country, racial and class-based ideas of deservedness have surrounded access to the technology. Though dialysis was made more available in the 1970s, racial, class, and geographical discrimination remains today. Following this section, the chapter focuses on the rise of for-profit dialysis companies that continue to influence how marginalized groups access dialysis. Since the 1990s, two dialysis companies – DaVita and Fresenius – have dominated the market share of dialysis outpatient centres in the United States, with results that have reinforced health disparities for patient consumers. Medical and health consumerism, or the ability to choose healthcare services, is a defining feature of many modern societies. Yet both 'patient' and 'consumer' have always been contested terms and limited by broader questions of race and racism, gender, class, and region. Where patients receive dialysis, whether in for-profit or not-for-profit clinics, in urban or suburban settings, or in rural spaces, influences both the quality of care that patients receive and their access to transplant lists. Emphasizing the experience of dialysis patients in the South, this second section makes the point that federal legislation of the 1970s did not single-handedly erase barriers to this important life-sustaining technology; rather, it changed the context to one in which access was more clearly shaped by the racism of geography and place.

Dialysis also speaks to a long history of race-based medicine expressed through and inextricable from technology.[5] By the 1990s, nephrologists had begun using race as a criterion for diagnosing

CKD on the misguided assumption that Black people universally have higher serum creatinine levels that should require a higher threshold of diagnosis. This chapter thus builds upon historical and social scientific research on technologies like the spirometer, pointing to yet another example of the deep and insidious interconnections of racism, chronic disease, technology, and patient consumerism that serves to both harm and deprive many Black Americans of their right to health. The chapter's conclusion points to a global history of CKD, technological failure, and the limits of dialysis technologies moving forward. In a volume on the consumer and patient, it addresses both in particular ways. Limited access to dialysis in certain geographical spaces restricted the consumerism of patients who had little choice on whether to use for-profit or not-for-profit dialysis. On the other hand, the continuation of misguided ideas and scientific racism from centuries ago play an important role in the diagnosis of kidney disease and treatment possibilities. At the centre of this story is dialysis, a medical technology often ignored in the history of medicine and technology scholarship. Dialysis does not provide a cure for kidney disease, yet it serves as a window into the evolution of patienthood and consumerism from the 1960s to the present. From an expensive technology in the 1960s, available to a limited number of people, to access being defined through a lens of disability in the 1970s and beyond, this technology continues to reify societal fissures and geographical racism. As with other forms of technology, dialysis tells us something about the spaces and worlds in which we have resided and continue to live, about lives valued and devalued, and about chronic disease and the limitations of medical technology.

Early history of dialysis

In 1944, after seventeen previous attempts, Willem Kolff successfully prolonged the life of a patient in renal failure with an artificial kidney in Holland. The emergence of nephrology in the 1940s as a medical speciality led to an increasing focus on kidney failure and diseases that involve the kidneys, such as diabetes and hypertension. In the United States, dialysis was first performed at the Peter Bent Brigham Hospital in Boston using the Brigham-Kolff dialyser. Expensive, time-consuming, and requiring an army of nurses, physicians, and technicians to operate, the Kolff dialyser was sparingly used by hospitals. Other hospitals stored the clunky machine in the closet where it collected dust. Very few consumers, mostly White and wealthy, could afford dialysis for an extended period. Moreover, because of complications that hampered its use, the Kolff dialyser was often considered a short-term intervention rather than a long-term option for renal failure.[6]

One of the early figures in the history of dialysis, physician Belding Scribner, first became interested in the function of kidneys in the 1940s while a medical student at Stanford University. Scribner was heavily influenced by the work of Thomas Addis, the most well-known kidney physician at the time. During his third year in medical school, he became fascinated with questions of fluid balance and potassium build-up after one of his patients died from kidney malfunction. While completing a residency at the Mayo Clinic in Rochester, Minnesota, in 1949, Scribner attended a lecture by Dr John Merrill of Boston who had begun working on the artificial dialysis. Scribner's initial interest in dialysis was not to prolong life but as a way of researching bicarbonate potassium and sodium intake.[7] Following residency, Scribner moved to Seattle, Washington, and began working at the Veteran's Administration Hospital and as faculty at the University of Washington School of Medicine. He spent the remainder of his career in Seattle where he was part of a team that created the AV shunt, allowing for multiple dialysis sessions while lowering infections and the complications that hampered earlier dialysis treatments. Scribner's creation of the shunt was one of the most important medico-technological innovations of the twentieth century, laying groundwork that would save millions of lives. Clyde Shields of Seattle was the first patient to receive the shunt in March 1960, and over the span of ten years would become the first chronic dialysis patient.[8]

Nevertheless, dialysis remained a rare and expensive technology. In the early 1960s, it was estimated that only one in three hundred people who needed it nationwide had access to dialysis and the cost was roughly $10,000 per person per year.[9] In January 1962 Scribner and other physicians in Seattle opened the first outpatient dialysis clinic in the nation, the Seattle Artificial Kidney Center (SAKC) at Swedish Hospital. From the outset, SAKC leaders confronted the ethical problems created by a life-sustaining technology limited in its availability. As costs prevented universal access, the SAKC developed medical, psychological, economic, and social criteria regarding who would be granted access to dialysis, and therefore the chance at life, and who would be left to die. An anonymous 'Admissions and Policy Committee' made up of local community members in Seattle made recommendations as to who would receive dialysis. The ethical ramifications of this were far-reaching.[10] The SAKC, in fact, created interest in the burgeoning field of medical ethics in the 1970s, becoming a prominent case study in undergraduate, professional, and medical courses. At the core were discussions that remain relevant today: 'who should get access to scarce health care goods and services, and at what price?'[11]

Candidates for dialysis at the SAKC had to meet certain age and health requirements. They could not be younger than fifteen or sixteen years of age or older than fifty, and could not have underlying medical conditions like diabetes or high blood pressure. The parameters disqualified many potential candidates.[12] An important turning point occurred when a young White girl, Caroline Helms, was diagnosed with CKD in 1964. She would not survive long without treatment. The SAKC Committee denied her dialysis, stating that she was too young for the harsh rigours of treatment. The Committee's denial of Helms led to the creation of the country's first home dialysis unit. Though many candidates, White and Black, young, and older, died in need of dialysis, a convergence of factors separated Helms from other patients. Much was made of the fact that Helms was an honour's student at a local Seattle high school, the type of person who, in the language employed at the time of worthiness and unworthiness, was deserving of a life-saving medical treatment. The language was veiled and there was no direct mentioning of her race, yet the narrative clearly framed her background and social status in implicitly racialized terms as a person whose death would be unacceptable in a civilized society.

Social theorist Achille Mbembe defines this as 'necropolitics', pitting those people who deserve protection against those who are expected to die.[13] The death of the powerless is expected. It is the normal consequences of a racialized and hierarchal society by those in power. Mbembe extends Foucault's theory of biopower or biopolitics, which Foucault defines as the sovereign control of individuals and populations through relations designed to let some live and others die within parameters that include reproduction and birth, allowance of death, sanitation, and disease.[14] Mbembe's necropolitics makes clear the legal sanctioning of violence and warfare that creates the 'living dead' and 'death worlds' for the powerless. Such forms of power include both bodily violence and forms of warfare where access to information, technology, drugs, people, and scarce resources is made available only to those deemed worthy of protection. Within this context, the death of certain individuals is rendered unacceptable and others justifiable, if not required, for maintaining White supremacy.[15]

Helms, a young White woman in a society that placed the most value on those interconnected identities, evoked sympathy and protection in ways not afforded to others. Further, the Helms family was able to elicit sympathy because they were well connected in the community with ties to Scribner and other Seattle physicians. Physicians in Boston and London had already transported dialysis tanks for temporary home use, but the process was costly and cumbersome. Scribner was distressed about Helms' imminent death and wrote to a friend asking if a miniature version of the current outpatient dialysis unit could be built so that patients might be dialysed overnight at home while sleeping. It would need to be constructed and then make its way through clinical trials and production within five months.[16] Money from the Hartford Foundation was made available, and a new team of engineers and clinicians built the 'Mini-I' for home use. The original outpatient dialysis unit, large and foreboding with tanks and wires, was known as the 'monster'. The home 'Mini-I' was a miniature version of the monster. Less than six months after Scribner's request, the 'Mini-I' was delivered to the SAKC where Helms and her mother were trained on using home dialysis, underscoring the resources and privileges that inform the history of technology as a form of access, social, and political power. Helms, and those who followed her, would receive dialysis twice per week for a ten-hour period overnight. The 'Mini-I' was more cost-effective than outpatient dialysis, and by 1965, increased production of the 'Mini-I' began to make home dialysis an option for more residents in Seattle.[17] Though home dialysis did not require going through a committee, the need for resources and financial stability remained necessary in ways that also made the 'Mini-I' exclusionary.[18]

Access to kidney dialysis was changed (at least theoretically) in October 1972 when Congress included access to dialysis treatment in a Social Security omnibus bill under the new Medicare law. This legislation did not emerge in a vacuum. After 1965, congressional bills were introduced every year to include dialysis under federal funding. Furthermore, broader debates on health insurance significantly influenced federal debates over dialysis. Throughout the country, kidney failure and dialysis were slowly making their way into the national consciousness of the late 1960s. The field of nephrology was strengthened with the emergence of the American Society of Nephrology by 1966, and amid the Civil Rights Movement in which health activists fought for increased visibility and awareness of long-neglected diseases impacting Black and non-White groups. The country was becoming increasingly attuned to kidney failure as a recognizable chronic disease that needed to be acknowledged. In the words of one scholar, 'As the number of physicians, treatment facilities, and patients increased, so too did local newspaper and television coverage of dialysis and kidney transplantation. In turn, members of Congress heard from and about constituents who needed dialysis ... The foundation for political action was being laid.'[19]

Dialysis benefited from the robust discussion over a national health insurance programme in 1971, spearheaded by Sen. Edward Kennedy (D-MA). The National Association of Patients on Hemodialysis (NAPH) was founded in 1969 by six dialysis patients to provide resources and community to those on dialysis and elevate the importance and seriousness of dialysis within broader healthcare politics. Shep Glazer was vice-president of the NAPH, and among other witnesses provided powerful testimony before the House Ways and Means Committee in November 1971. 'I am 43 years old, married for 20 years, with two children, ages 14 and 10. I was a salesman until a couple of months ago until it became necessary for me to supplement my income to pay for the dialysis supplies,' Glazer testified. 'I tried to sell a noncompetitive line, was found out, and was fired. Gentlemen, what should I do? End it all and die? ... Please tell me. If your kidneys failed tomorrow, wouldn't you want the opportunity to live? Wouldn't you want to see your children grow up?' In a controversial move, Glazer underwent a dialysis procedure on the US House Floor as part of his testimony so that lawmakers could visualize the workings of dialysis. It was later acknowledged that Glazer's session was terminated after just five minutes when the physician attending the session determined he was suffering from ventricular tachycardia, a condition in which the chambers of the heart start to beat abnormally quickly. Nonetheless, the point was made.[20]

The crux was defining the need for chronic dialysis treatment as a disability under the Medicare programme. Part of President Lyndon Johnson's Great Society, Medicare and Medicaid were conceptualized differently from the outset. Medicaid was a programme built on matching state and federal dollars to provide healthcare for women and poor people. Medicare was widely celebrated at the time as an 'entitlement' and the 'single most important expansion of health care in the twentieth century'. Rather than the belief by some people that Medicaid was a 'poor person's program' for an undeserving population, Medicare was lauded as providing medical care to worthy and deserving elderly citizens.[21] The 1972 Social Security Amendment provided the entry point for more than 90 per cent of people needing dialysis to be covered under Medicaid and Medicare.[22]

CKD was reshaped as a disability, but focus was also placed on the 'chronic' aspect of the disease. As historian George Weisz has written, historians of medicine have only recently begun defining the complicated parameters of chronic disease in comparison to the more robust scholarship on epidemics.[23] Important scholarship on sickle cell disease, cancer, diabetes, and heart disease signal vital new directions on the ways in which poor housing, poverty, structural racism, stigma, and medical discrimination fuel chronic disease disparities in the historical past and present.[24] Similarly, historian Beth Linker pushes historians of medicine to pay closer attention to the ways in which disease produces disability, or how disabilities are complicated by the mechanisms of disease.[25] Inclusion of dialysis under the Social Security Amendments created the parameters for a new chronic patient identity by the 1970s, known as ESRD. ESRD is the final stage of progressing chronic kidney failure that requires either dialysis or a transplant for the patient to live. The Social Security Amendments did not specifically address racial disparities in access to dialysis treatment. Yet, the categorization of ESRD, disability, and chronic disease converged in a new patient identity that served to both provide access to dialysis while simultaneously highlighting long-standing health disparities.[26]

Black newspapers highlighted stories of vulnerability and lack of access in the years following the legislation. Blondell Williams was a Black woman who had worked for years as a practical nurse at Chicago's Cook County Hospital in 1973 when she suddenly gained thirty-four pounds in just three weeks. She visited her private physician and was admitted to the hospital for a month, after which she was informed that she had an incurable kidney issue. Other physicians at the hospital reinforced the seriousness of her kidney disease and agreed there was nothing else that could be done for her.[27] The reasons why dialysis and transplantation were not initially suggested are unclear but are indicative of the ways that dialysis law continued to reify deep social chasms in American society. It is hard to believe, for instance, that dialysis and transplantation would not have been the first option for a middle-class White woman living in a wealthy neighbourhood. Williams, a mother of four, took things into her own hands and called a kidney specialist team at Chicago's Billings Hospital; upon hearing her recent medical history, they suggested she check herself into the hospital as soon as possible where she was immediately placed on dialysis three times per week.[28]

Other examples of kidney failure among Black people in Chicago highlight the seriousness and immediate need for life-saving treatment in the wake of the dialysis law. In the same year, Robert Taylor began suffering from bouts of extreme nausea, swelling of joints, and dizziness at the age of forty-seven. Initially, he assumed his body was fighting off a severe cold until he passed out after sneezing and woke up in the hospital with water accumulated around both legs. His kidneys had failed, and unlike Williams, he was immediately referred to Billings Hospital where he received medication for ninety days and was placed on dialysis for several months. Taylor was so overjoyed when receiving news of a kidney transplant that he immediately left home and arrived at the hospital in pyjamas. Sylvia Marshall would also receive a kidney transplant in the Chicago area at the age of thirteen. She was just five years old when she began complaining to her mother of tiredness after a Sunday school picnic. Her body was also swelling from fluid build-up and her mother took her to the hospital. She was placed on dialysis after both of her kidneys were removed and, as a child, had to manage dialysis three times per week amid school obligations and the many desires that children have.[29]

Dialysis and disability are complicated by questions of racism and anti-Blackness. As the story of Blondell Williams shows, federal legislation by itself does not guarantee access to dialysis when other forms of bias are present. Beyond the 1970s and to the present day, access to dialysis has been influenced by more than legislation; it has been shaped by the conditions of the states and communities in which people reside and whether dialysis occurs in not-for-profit hospital settings or for-profit clinics. The 1972 dialysis legislation was not specifically intended to erase racial bias in the application of dialysis. Stories printed in Black newspapers revealed a long-neglected population suffering from CKD, showing the ways in which persistent forms of racism and bias have continued to render Black pain invisible. The remainder of this chapter pivots on the story of Williams and the ways in which access to dialysis remains limited for many reasons.

Rise of a duopoly and for-profit dialysis?

Today, the South has among the highest rates of CKD in the country. Low wages, high unemployment, significant under- and uninsured populations, rural or urban food, and medical deserts cause parts of the Southern United States to be among the unhealthiest in the United States. Rates of Black people suffering from poverty-related diseases such as diabetes, hypertension, stroke, and ESRD remain high in the South. Parts of the South – including Florida, Alabama, Mississippi, Louisiana, and Texas – have been known as the 'stroke belt', a term used to define the preponderance of stroke and hypertension disorders in the region since the 1940s.[30] Medical sociologist Anthony Ryan Hatch showed that, by the 1970s, the importance of a constellation of specific clinical and laboratory measurements – elevated blood pressure, cholesterol, blood pressure, and weight – was framed as metabolic syndrome. This was not so much a 'disease' as what was considered a disease in the making.[31] Elevated levels of these four key measurements were considered a window into potential risks for diabetes, stroke, heart disease, and other conditions. Hatch also pointed out that such evaluations were premised on the repackaging of old biological ideas of racial difference:

> metabolic syndrome draws upon and extends knowledge-making practices that have long constructed race as natural, biological, and genetic. As the biomedical discourses and practices of metabolic syndrome continue to unfold, they intersect with the ways in which race shapes the theories and practices of medicine in terms of disease surveillance, diagnosis, and treatment.[32]

Since the 1990s, the surveillance, diagnosis, and treatment of CKD and ESRD have evolved alongside the emergence of a for-profit dialysis industry dominated by two major companies: DaVita and Fresenius. The for-profit dialysis industry influences patient care, treatment, and, most importantly, access to transplants. DaVita Kidney Care began as Medical Ambulatory Care in 1979, and after an acquisition in 1994 by DLJ Merchant Banking Partners, the name changed to Total Renal Care Holdings. The name was again changed to DaVita in 2000 and is now known as DaVita Kidney Care. In 1996 Fresenius SE and Company dialysis unit merged with National Medical Care to form Fresenius Medical Care. With headquarters in Bad Homburg vor der Hoehe, Germany, and Waltham, Massachusetts, Fresenius is the global leader of dialysis in terms of revenue and patients, and has a production arm that manufactures dialysis equipment.[33]

Statistics tell the tale of the stranglehold these two companies have in the industry. By 2019 Fresenius served roughly 208,000 dialysis patients in the United States, and DaVita about 204,000. The number of patients served by DaVita was eight times more than the next dialysis provider on the list, US Renal Care, which served 25,000 patients.[34] Fresenius operates approximately 2,600 outpatient clinics in the United States and 3,900 clinics throughout the world with over 330,000 patients, over half in the United States. DaVita operates close to 2,800 clinics in the United States. Together, DaVita and Fresenius hold around 80 per cent of the United States dialysis industry market share. As one scholar wrote recently, when it comes to dialysis, most people have a choice between either Coke or Pepsi.[35]

These numbers, however, only tell half the story. Depending on their zip code and region of the country in which they live, many people find it difficult to access dialysis at all. Southern states consistently have the highest percentage of kidney failure and dialysis in the United States. Both are the result of poverty in the region, poor nutrition, environmental stressors such as living in food deserts, and having among the highest per capita cases of chronic disease in the country. All have converged into a public health emergency of CKD and dialysis in the region.[36] Alabama and Mississippi are on the top of this list that also includes Tennessee, Georgia, Louisiana, North Carolina, and South Carolina. Recent Medicare statistics show that most dialysis clinics in the United States are for profit, and that better care is associated with not-for-profit clinics such as publicly funded hospitals and even the much-maligned VA hospital system.[37] The trend takes on even more importance in Southern states where there exists a long history of medical racism and structural inequality. A 2016 report on dialysis in Southern states suggested that of the 1,241 clinics in Alabama, Georgia, Louisiana, Mississippi, and the Carolinas, only 104 were not-for-profit. The report also suggested a harsh rural/urban and racial divide as most not-for-profit clinics operated in overwhelmingly White, urban, and affluent areas. Access to either for-profit or not-for-profit clinics is also limited in rural geographies, though the for-profit industry has consistently targeted poor, rural, and predominately Black and minority communities. There are far more rural counties in these states without access to a single dialysis clinic, an important variable as proximity to dialysis clinics has been linked with lower mortality rates.[38]

Accessibility to dialysis must also be understood within a differential experience of chronic disease. Roughly 75 per cent of people in the United States are unemployed on the day they begin dialysis treatment. Journalist Laura Abraham wrote in her 1993 seminal work, *Mama Might Be Better off Dead: The Failure of Health Care in Urban America*: 'Dialysis itself is easily a part-time job … counting travel, waiting, and treatment time, the process can eat up six hours a day, three times a week.'[39] Unlike with some other chronic diseases, like breast cancer for instance, we are not bombarded with images of dialysis patients with the yearly awareness campaigns. Kidney failure remains a constant, if invisible, source of chronic social and bodily pain. Those exposed to a mostly for-profit dialysis clinic environment, living in geographically marginalized spaces, and residing in poorer countries across the globe, have higher mortality rates. Patients and family are often left to advocate for themselves and frequently complain about poorly trained staff, lack of on-duty physicians, and dialysis-related chronic needs that remain unaddressed. Broken furniture and blood splattered on chairs, floors, and walls are common in poorly run clinics.[40] Beyond being unsightly and unsanitary, such conditions might also incur medical complications and infections. Dialysis clinics have been suspended or closed because of cross-infection that resulted in patients or staff being exposed to hepatitis C, tuberculosis, and HIV.[41] As a 2010 *Atlantic* article made painfully clear, mistakes in prescriptions have resulted in the death of patients. A particularly egregious example occurred in 2005. Thirty-nine-year-old Henry Baer had recently been placed on dialysis in Prescott Valley, Arizona, the result of uncontrolled diabetes and high blood pressure. During his third session on New Year's Eve, Baer's bloodline was accidentally disconnected. After seeing blood spraying everywhere, the technicians on duty panicked and quickly reconnected the catheter and line without adhering to required emergency standing orders for reduced contamination. Not long after leaving the facility, Baer began complaining of nausea and started convulsing. He was suffering from an antibiotic-resistant staph infection from his catheter that quickly spread to his heart and brain. Baer died just a few days after receiving dialysis.[42]

For profit dialysis clinics are also shaped by the larger forces of the labour market. Full-time physicians and nurses can be difficult to retain for clinics. Technicians can begin working with a high school diploma and the passing of a certification test. Though many of these technicians perform duties to the best of their ability, the lack of consistent training can make for a difficult, frustrating, and potentially deadly medical environment. Many patients have little recourse in criticizing for-profit clinics, particularly if a racialized geography makes it difficult for them to find another clinic.[43] Historian Steven Peitzman and others have coined the term of 'Dear John' letters sent to vocal and activist dialysis patients considered troublemakers by clinic administrators and staff. Patients might check mailboxes or receive an email one day to discover a curt notice stating without reason that dialysis services for them are being discontinued.[44]

While a transplant is the best possible treatment for people with kidney failure, there remain significant barriers for poor and minority patients exposed to the for-profit dialysis industry. Most people will require a deceased donor transplant, though less than 14 per cent of the roughly half a million dialysis patients have managed to make their way on to the waiting list. A 2019 article in the *Journal of the American Medical Association (JAMA)* put it in clear terms: 'receiving dialysis at for-profit facilities compared with non-profit facilities was associated with a lower likelihood of accessing kidney transplantation'.[45] The Centers for Medicare and Medicaid Services mandate that information and education regarding kidney transplants, and the protocol for enrolment on a transplant waiting list, must come from active coordination between patients, social workers, and physicians at dialysis clinics. The for-profit clinic referral system is also ripe with neglect in this regard. 'For-profit dialysis facilities have a lower standardized transplantation ratio, and their patients are less likely to be waitlisted compared with nonprofit facilities. Physicians at for-profit dialysis facilities are less likely to have detailed discussions with patients about transplantation or involve families in the discussion', according to the *JAMA* study.[46] Medical writer Liz Presser also highlights the ways in which lack of insurance, cracks in the referral protocols, and confusing information regarding requirements and qualifications for transplant waitlists frustrate dialysis patients. Black people are at a significant disadvantage when it comes to medical care at every stage leading to transplants.[47] In the twelve months leading to the start of dialysis, Black people are 18 per cent less likely to be evaluated by a nephrologist, meaning that care is delayed to the point of a medical emergency that cannot be ignored, and dialysis is the only option to prevent a potentially deadly toxic build-up of waste in the bloodstream.[48]

In Southern rural counties, access to specialists including nephrologists, vascular surgeons, ophthalmologists, podiatrists, and oncologists remains an issue alongside dialysis. Fewer hospitals and clinics capable of providing vascular surgery to improve the blood flow of people suffering from diabetes have led to an epidemic of limb amputations in Mississippi and other Southern states. Thus, the for-profit dialysis industry is just one symptom of the broader failure of healthcare in the South and the neglect of poor, Black, and minority dialysis patients.[49]

Technology, eGFR, and race-based medicine

Historians and theorists have long considered the human dimensions of technology, particularly the ways in which technology has been used to divide, conquer, and suppress individuals and groups, as well as reify social and politically constructed difference.[50] Langdon Winner asked the question in 1997: 'Do Artifacts Have Politics?', influencing a generation of scholarship on the meaning of technology in society and vice versa. 'At issue', Winner writes, 'is the claim that the machines, structures, and systems of modern material culture can be accurately judged not only for their contributions to efficiency and productivity ... but also for the ways in which they can embody specific forms of power and authority'.[51] In his 1999 work, *Drawing Blood: Technology and Disease Identity in Twentieth-Century America*, historian of medicine Keith Wailoo similarly shows the ways in which technology used to diagnose and treat disease does not operate independent of a broader social and political discourse. Wailoo shows that there is not just one way to read an EKG, for instance, but multiple ways. How physicians decode an EKG can lead to the diagnosis or elimination of heart disease as a medical opinion with important ramifications for the patient. Assumptions of biology, race, class, gender, and other factors thus influence how physicians interpret EKGs and the creation of a disease identity for patients. This identity is important not only for diagnosing disease but also for the acknowledgement of disease and treatment of suffering. Technologies can also reinscribe a history of neglect and obscure experiences of marginalized groups. 'Depending upon its user, its interpreters, and the context,' Wailoo writes, '... technology can have both oppressive and liberating effects'. Physicians and other medical professionals define and give meaning to technology.[52]

Historian and theorist Lundy Braun emphasizes the connection between plantation-era science and current-day genetics through the spirometer in ways that have perpetuated long-debunked theories of Black biological difference and inferiority.[53] Dating back to the nineteenth century, the spirometer technology measured lung functioning and output and was interpreted by physicians during the antebellum and post-bellum era to justify slavery and subjugation of Black bodies. The spirometer was interpreted by pro-slavery and anti-Reconstruction physicians to suggest that Black people had a lower lung output and 'vital capacity' than Whites. Following the Civil War, the idea of a lower lung capacity was used as justification for continued racial violence and repressive laws under the banner that recently freed Black people lacked the stamina and physiological vitality to endure the rigours of freedom, citizenship, and democracy. Well over a century later, racial bias is built into the design of the technology itself. The spirometer is still in use today, and it is not uncommon for clinicians to employ a race correction function that assumes biological differences between Black and White normal lung output.[54]

Even more recent and closer to questions of patient consumerism is the contentious history of BiDil, approved in 2005 by the Food & Drug Administration (FDA) as the first so-called race-specific drug technology. A combination of two drugs, isosorbide dinitrate and hydralazine hydrochloride, BiDil was marketed to reduce congestive heart failure in people who self-identified as Black or African American. The identification of BiDil as a race-specific drug was problematic from the outset as there was no evidence that the drugs worked only in one specific population group. In other words, the combination of these drugs proved as effective in people who self-identified as White and there was no evidence the drugs worked specifically, or solely, in Black people. The two generic drugs were resurrected, repackaged, and marketed to target Black people disproportionately suffering from congestive heart failure.[55] Legal, ethics, and social science scholars were critical of the use of race in both the clinical trials and the suggestion of therapeutic value. What does it mean to self-identity as 'Black', and how do you measure biological Blackness in clinical trials? Moreover, what does it mean if a person has one self-identified Black and one self-identified White parent? BiDil initially failed FDA approval for general use due to insufficient analysis, data, and contradictions over efficacy. Yet it would reappear as a drug to somehow reduce racial health disparities. The creators could never clearly articulate what was so unique about Black bodies that were so receptive to BiDil. Clinical studies showed that BiDil was no more efficacious in self-identified Black people than in any other ethnic group.[56] BiDil also reignited debates over the meaning of race and biological difference between various groups. The idea that drug technology could be developed and marketed to reduce health disparities was fallacious, dangerous, and undermined the work around social determinants of health. If a drug could be developed to address hypertension and heart disease among Black people, the social environmental factors of poor housing, lack of medical insurance, unemployment, food deserts, and the daily stress of systematic and chronic racism could be wilfully ignored.

Dialysis is also part of this recent history of constructing biological differences through so-called race. Since the 1990s, physicians have often diagnosed CKD by using an estimated glomerular filtration rate (eGFR) but do so by employing the practice of race-adjusted algorithms. Levels of serum creatinine are measured to determine eGFR for kidneys. Higher eGFR values are correlated with better kidney functioning. Algorithms are routinely race-corrected or adjusted so that self-described or otherwise identified Black people are routinely assigned higher values. An eGFR score of twenty is normally required for diagnosing kidney failure among all groups. Yet, even before other diagnostic criteria are observed, the eGFR score for Black people can be increased by one to two points or more, resulting in a higher bar for the diagnosis of kidney failure in comparison to White people.[57] The claims from developers of race-based eGFR algorithms suggest that Black people have higher serum creatinine levels than Whites. A recent *New England Journal of Medicine* essay addressed the spurious claims of supposed biological differences around eGFR, arguing that 'Explanations that have been given for this finding include the notion that black people release more creatinine into their blood at baseline, in part because they are reportedly more muscular. Analyses have cast doubt on this claim, but the "race-corrected" eGFR remains the standard.'[58] What being 'more muscular' means is remarkably void of scientific scrutiny and analysis, and in the words of Barbara and Karen Fields, such uses of race require a jump in logic and 'half-lit zone of the mind's eye'.[59] Unthinkingly lumping all so-called Black people into a single monolithic and pre-assumed group clearly links scientific racism of the past with the present.

The results of this algorithm-based racial discrimination are devastating. Some reports suggest that as many as one million Black people might be treated earlier if the correction were removed.[60] The eGFR score can influence access to clinical trials, lead to misdiagnosis and ill-informed drug treatment regimens, delay referral to nephrologists and other specialists, result in an unchecked progression of kidney failure, and, importantly, prevent referral to a kidney transplant waiting list. As described earlier in this chapter, unclear standards of referral from the major dialysis providers DaVita and Fresenius remain a significant barrier for Black people. eGFR race-based algorithms represent yet an additional burden to accessing scarce kidney resources.[61] Furthermore, race adjustments are used in determining eligibility for kidney transplants. Donor kidneys coming from bodies identified as Black are often provided with a higher kidney donor risk index (KDRI) score, suggesting a much higher risk of graft transplant failure to a recipient. Though scientists fail to explain the specific race-based causes for this disparity, KDRI scores lower the availability of Black donors' kidneys and contribute to the much longer transplant wait times (four times as long) for Black people who, like everyone else, look to relatives as a first option for transplants. KDRI scores can make donations from family relatives more difficult. For Black people in particular, their difficulty in being diagnosed with CKD and the notoriously long wait times for anonymous donations on national kidney lists remain significant barriers.[62]

It is ironic that since the 1970s, claims of higher-than-average rates of ESRD and CKD among Black people have become normalized in scientific literature. Yet, misguided and racialized ideas of Black biological difference in the form of eGFR rates have resulted in higher barriers for diagnosis in ways that continue to render the pain of chronic disease among Black people invisible. What do we make of this historical paradox? Calls for ending the use of eGFR have come from *JAMA*, and a few medical centres, including Beth Israel Deaconess Medical Center in Boston, the University of Maryland, and the University of Washington Medical Center in Seattle, recently discontinued its use. On a grassroots level, medical students, influenced by the rise of medical humanities training in undergraduate and professional curriculum, and responding to the Black Lives Matter Movement, have also been vocal in calling for an end to the practice.[63] Yet there are those who hold firm to the validity of eGFR or suggest that it should remain in use until something better comes along. The long history of racial science makes clear that dubious knowledge and understanding can more easily lead to the implementation of harmful policies like eGFR, but reversing such policies predicated on faulty premises and failing to take into consideration systemic racism in American society are difficult. The future of eGFR will remain a contested battlefield.[64]

Global futures of technological failure

Across the globe, dialysis rates have been increasing in Africa, Latin America, and southern Asia, particularly India. One recent report suggested that dialysis was available in thirty-four African countries, though access remains a significant issue for most of the continent.[65] The lack of usable water and electricity is a persistent issue for dialysis in poorer countries. Kidney failure has recently led to awareness of a novel condition that seems to disproportionately impact workers in particular regions of the world. What has been called Meso-American chronic kidney disease of unknown origin or Meso-American nephropathy was defined by clinicians beginning in the 1990s.[66] Experts noticed a steep uptake of ESRD among sugar, cotton, and corn agricultural workers in the Pacific Ocean Central American countries of El Salvador, Guatemala, Nicaragua, Panama, and Costa Rica. Related forms of this kidney failure have also been diagnosed among agricultural workers in Sri Lanka and India. It was striking to clinicians that diagnoses of ESRD did not stem from the usual range of co-factors that included diabetes, hypertension, and glomerular disease. In many cases, elevated serum levels were discovered before the harvest season. What was causing kidney failure to occur in such vulnerable populations? Researchers have pivoted on three potential causes for this unique form of kidney failure that still impacts agricultural workers today: an unknown infectious agent, exposure to pesticides in agricultural working communities, and heat-related injury to the kidneys.[67]

Researchers have focused on the widespread toxicity of agro-chemicals such as glyphosate that might expose workers and adjacent communities via contaminated water supplies to harmful neurotoxins. Exposure to airborne lead and silicosis particles and infectious zoonotic diseases such as leptospirosis and hantavirus, spread via domestic animals, farm animals, and rodents, have all demanded attention. Yet, heat-related kidney injuries among sugarcane workers labouring in low-lying areas received the most sustained attention. 'One striking finding is that the regions in which chronic kidney disease has been reported tend to be the hottest regions in the various countries,' wrote researchers in a recent essay.[68] Sugarcane workers at lower levels seem to be at more risk than those at elevated levels where temperatures are cooler. The exploitation of seasonal agricultural labourers working twelve to sixteen hours in excessive heat, and little access to drinkable water, have led to questions of whether heat stress might be the cause of kidney injuries. Global warming as an evolving health and environmental justice issue will only exacerbate these racial and socioeconomic disparities in the future. Whether infection-based, the result of overexposure to pesticides or heat, or some combination of these or other causes, this novel illness that impacts kidneys is altering the ways in which we must think about kidney failure and dialysis in the future.[69]

Finally, the future of dialysis is driven by questions of technological innovation and telehealth, both of which have the potential to further divide dialysis patients. Researchers have begun working on implantable artificial kidneys to replace traditional dialysis. Other innovations include portable kidney technology that improves vascular access and simulation of actual kidney functioning. The hope is that miniature technologies and portable dialysis will free those with ESRD from the need to travel to a clinic three times per week.[70]

Yet, dialysis is evolving alongside broader ethical issues within telemedicine. Dialysis patients who can provide trackable data and receive clinical changes to treatment remotely will have the potential to increase their quality of life. Though it has roots in the 1950s and 1960s, the COVID-19 pandemic has strengthened the call for telemedicine.[71] The origins of telemedicine as a community-based tool employed to connect physicians and nurses with neglected populations have also been reshaped in more recent decades by private companies in ways that reproduce the digital divide. In the current age of technological apps and platforms for health systems, doubts about the quality of care for some patients are exacerbated by issues of not owning computers, tablets, and smart telephones needed to access telemedicine. Though the future of dialysis seems ripe for significant technological innovation, the inherent problem is whether this progress will lead to beneficial changes for everyone or serve to further reinscribe differences along racial and socioeconomic lines that have defined this complicated and life-saving medical technology since the 1960s.[72]

This chapter has defined some of those racial and socioeconomic differences in access to dialysis from the SAKC, to the rise of dialysis companies, and finally to problematic race-based eGFR evaluations in the diagnosis of kidney disease. At the core of these discussions are patient consumers. Since the 1960s, recipients of dialysis have been either patients or consumers, sometimes both, and at other times one in search of the other. Dialysis has been the vehicle through which these conversations and debates have raged. This important medical innovation has saved millions of lives and transformed ESRD into a chronic disease. Yet technology always has politics. Dialysis has continuously proven to be a window into broader questions of societal worth, geographical racism, and long-standing tenets of scientific racism towards Black bodies. Looking towards the future of telemedicine, technological changes to dialysis are already deepening chasms of race- and class-based access to medicine based on insurance and the ability to pay. Historians tell stories in time periods. Yet, there are moments when comparisons are informative and telling. The story of dialysis in the 1960s and today reveals a technology that both patients and consumers alike still find problematic and painfully difficult to access.

Notes

1 Caroline Hsu and Daniel Weiner, 'Covid-19 in Dialysis Patients: Outlasting and Outsmarting a Pandemic,' *Kidney International* 98, no. 6 (December 2020): 1402–4; National Kidney Foundation, 'Kidney Disease: The Basics.' www.kidney.org/news/newsroom/fac tsheets/KidneyDiseaseBasics.
2 www.niddk.nih.gov/health-information/kidney-disease/kidney-fail ure/hemodialysis. This essay primarily focuses on haemodialysis.
3 www.ncbi.nlm.nih.gov/books/NBK492981/.
4 Marcello Tonelli, Raymond Vanholder, and Jonathan Himmelfarb, 'Health Policy for Dialysis Care in Canada and the United States.' *Clinical Journal of the American Society of Nephrology* 15, November 2020: 1669–80. ESRD is considered the final stage of CKD. CKD can

be reversible unless a diagnosis of ESRD occurs, at which point dialysis is required. For this essay, I will refer primarily to ESRD.

 5 Keith Wailoo, *Drawing Blood: Technology and Disease Identity in Twentieth-Century America* (Baltimore, MD: Johns Hopkins University Press, 1999); Lundy Braun, *Breathing Race into the Machine: The Surprising Career of the Spirometer from Plantation to Genetics* (Minneapolis, MN: University of Minnesota Press, 2014).

 6 Steven Peitzman, *Dropsy, Dialysis, Transplant: A Short History of Failing Kidneys* (Baltimore, MD: Johns Hopkins University Press, 2007); Renee C. Fox and Judith P. Swazey, *The Courage to Fail: A Social View of Organ Transplants and Dialysis* (Chicago, IL: University of Chicago Press, 1978).

 7 Belding H. Scribner Interview, 26 March 1986, Belding H. Scribner Papers, Acc 3301–009, Tapes 1–4, University of Washington Special Collections.

 8 Scribner Interview.

 9 Scribner Interview; Michael Bliss, *The Discovery of Insulin* (Chicago, IL: University of Chicago Press, 1982).

10 Susan Reverby, *Examining Tuskegee: The Infamous Syphilis Study and Its Legacy* (Chapel Hill, NC: University of North Carolina Press, 2013); David Rothman, *Strangers at the Bedside: A History of How Law and Bioethics Transformed Medical Decision-Making* (New York: Basic Books, 1992).

11 Amy Gutmann and Jonathan Moreno, *Everybody Wants to Go to Heaven but Nobody Wants to Die: Bioethics and the Transformation of Health Care in America* (New York: Liveright Publishing, 2019), p. 123.

12 Jerry Pendras, 'Experience with Patient Selection for Chronic Hemodialysis.' Unpublished manuscript from the Seattle Artificial Kidney Center, Belding Scribner Papers, ACC 3301–82–14, Restricted Files, Box 3, University of Washington Libraries Special Collections.

13 Belding Scribner and Albert Babb, 'Chronic Hemodialysis in Seattle: 1960–1966.' unpublished manuscript, ACC 3301–05, Box 2, Belding Scribner Papers, University of Washington Libraries Special Collections, pp. 15–16; Achille Mbembe, *Necropolitics* (Durham, NC: Duke University Press, 2019).

14 Michel Foucault, *The Birth of Biopolitics: Lectures at the College de France, 1978–1979* (New York: Picador Press, 2010); Mbembe, *Necropolitics*.

15 Mbembe, *Necropolitics*.

16 Scribner and Babb, 'Chronic Hemodialysis', pp. 15–16.

17 Scribner and Babb, 'Chronic Hemodialysis', pp. 15–16.

18 Scribner and Babb, 'Chronic Hemodialysis', pp. 15–16.
19 Richard Rettig, 'Origins of the Medicare Kidney Disease Entitlement: The Social Security Amendments of 1972', in *Biomedical Politics*, edited by Kathi Hanna, Division of Health Sciences Policy, Committee to Study Biomedical Decision Making, and the Institute of Medicine, pp. 176–208 (Washington, DC: National Academies Press, 1991).
20 National Health Insurance Proposals: Hearings before the Committee on Ways and Means, House of Representatives, 92nd Congress (3–4 November); US Government Printing Office, Washington, DC, 1972; Testimony of Shep Glazer, 1539; Rettig, 'Origins of the Medicare Kidney Disease.'
21 Alan B. Cohen, David C. Colby, Keith A. Wailoo, and Julian E. Zelizer (eds), *Medicare and Medicaid at 50: America's Entitlement Programs in the Age of Affordable Care* (New York: Oxford University Press, 2015).
22 Rettig, 'Origins of the Medicare Kidney Disease.'
23 George Weisz, *Chronic Disease in the Twentieth Century: A History* (Baltimore, MD: Johns Hopkins University Press, 2014).
24 Keith Wailoo, *Dying in the City of the Blues: Sickle Cell Anemia and the Politics of Race and Health* (Chapel Hill, NC: University of North Carolina Press, 2001); Keith Wailoo, *How Cancer Crossed the Color Line* (New York: Oxford University Press, 2011); Anne Pollock, *Medicating Race: Heart Disease and Durable Preoccupations Difference* (Durham, NC: Duke University Press, 2012); Arleen Tuchman, *Diabetes: A History of Race and Disease* (New Haven, CT: Yale University Press, 2020).
25 Weisz, *Chronic Disease in the Twentieth Century*; Beth Linker, 'On the Borderland of Medical and Disability History: A Survey of the Fields.' *Bulletin of the History of Medicine* 87, no. 4 (Winter 2013): 499–535.
26 Peitzman, *Dropsy, Dialysis, Transplant.*
27 'Kidney Victims Find Faith.' *The Chicago Defender*, 17 November 1973.
28 'Kidney Victims Find Faith.'
29 'Kidney Victims Find Faith.'
30 Douglas Lanska and Lewis Kuller, 'The Geography of Stroke Mortality in the United States and the Concept of a Stroke Belt.' *Stroke* 26 (1995): 1145–9; Daniel Lackland and Michael Moore, 'Hypertension-Related Mortality and Morbidity in the Southeast.' *Southern Medical Journal* 90 (February 1997), 191–8.
31 Anthony Ryan Hatch, *Blood Sugar: Racial Pharmacology and Food Justice in Black America* (Minneapolis, MN: University of

Minnesota Press, 2016), pp. 10–11; Robert Aronowitz, *Making Sense of Illness: Science, Society, and Disease* (Cambridge: Cambridge University Press, 1998).

32 Hatch, *Blood Sugar*, pp. 10–11.

33 See https://healthcareappraisers.com/2020-outlook-dialysis-clinics-and-esrd/.

34 See www.statista.com/statistics/807627/largest-dialysis-providers-us-by-number-of-patients/.

35 https://healthcareappraisers.com/2020-outlook-dialysis-clinics-and-esrd/; Robin Fields, 'In Dialysis: Life Saving Care at Great Risk and Cost.' *ProPublica*, 9 November 2010.

36 Richard M. Mizelle, Jr, 'Diabetes, Race, and Amputations.' *The Lancet: Art of Medicine* 397, no. 10281 (April 2021): 1256–7.

37 'For-Profit vs. Nonprofit Dialysis Centers: Alabama, Georgia, Louisiana, Mississippi, North Carolina, and South Carolina AKA the Deep South.' 2016, Tufts University Report: https://sites.tufts.edu/gis/files/2020/08/couture_sara_ph262_Spring2020.pdf; Tonelli, Vanholder, and Himmelfarb, 'Health Policy for Dialysis Care in Canada and the United States.'

38 'For-Profit vs. Nonprofit Dialysis Centers.'

39 Jonathan Himmelfarb, Raymond Vanholder, Rajnish Mehrotra, and Marcello Tonelli, 'The Current and Future Landscape of Dialysis.' *Nature Reviews Nephrology* 16 (2020): 573–85; Laurie Kaye Abraham, *Mama Might Be Better Off Dead: The Failure of Health Care in Urban America* (Chicago, IL: University of Chicago Press, 1993), p. 39.

40 Robin Fields, 'God Help You. You're on Dialysis.' *The Atlantic*, December 2010; Peitzman, *Dropsy, Dialysis, Transplant*; and other essays.

41 Fields, 'God Help You.'

42 Fields, 'God Help You.'

43 Peitzman, *Dropsy, Dialysis, Transplant*.

44 Peitzman, *Dropsy, Dialysis, Transplant*.

45 Jennifer Gander, Xingyu Zhang, Katherine Ross, Adam Wilk, Laura McPherson, Teri Browne, Stephen Pastan, Elizabeth Walker, Zhensheng Wang, and Rachel Patzer, 'Association between Dialysis Facility Ownership and Access to Kidney Transplantation.' *JAMA Original Investigation* 322, no. 10 (2019): 957–73.

46 Gander et al., 'Association between Dialysis Facility Ownership.'

47 Liz Presser, 'Tethered to the Machine.' *ProPublica*, 15 December 2020.

48 Presser, 'Tethered to the Machine.'

49 Mizelle, Jr, 'Diabetes, Race, and Amputations.'

50 Daniel Headrick, *The Tools of Empire: Technology and European Imperialism in the Nineteenth Century* (New York: Oxford University Press, 1981); Michael Adas, *Dominance by Design: Technological Imperatives and America's Civilizing Mission* (Cambridge: Harvard University Press, 2006).

51 Langdon Winner, *The Whale and the Reactor: A Search for Limits in an Age of High Technology* (Chicago, IL: University of Chicago Press, 1989), p. 19.

52 Wailoo, *Drawing Blood*, p. 2.

53 Braun, *Breathing Race into the Machine*.

54 Braun, *Breathing Race into the Machine*.

55 Dorothy Roberts, 'What's Wrong with Race-Based Medicine? Genes, Drugs, and Health Disparities.' *Minnesota Journal of Law, Science & Technology* 12, no. 1 (Winter 2011): 1–21; Jonathan Kahn, *Race in a Bottle: The Story of BiDil and Racialized Medicine in a Post-Genomic Age* (New York: Columbia University Press, 2012).

56 Roberts, 'What's Wrong with Race-Based Medicine?'

57 Presser, 'Tethered to the Machine.'

58 Darshali Vyan, Leo Eisenstein, and David Jones, 'Hidden in Plain Sight – Reconsidering the Use of Race Correction in Clinical Algorithms.' *The New England Journal of Medicine* 383, no. 9 (August 2020): 874–82.

59 Karen Fields and Barbara Fields, *Racecraft: The South of Inequality in American Life* (New York: Verso Books, 2012).

60 Jyoti Madhusoodanan, 'Is a Biased Algorithm Delaying Health Care for Black People.' *Nature* 588 (December 2020): 546–7.

61 Vyan, Eisenstein, and Jones, 'Hidden in Plain Sight.'

62 Vyan, Eisenstein, and Jones, 'Hidden in Plain Sight.'

63 Richard M. Mizelle, Jr, 'Getting into Good Medical Trouble.' *Los Angeles Review of Books*, 6 November 2020.

64 Madhusoodanan, 'Is a Biased Algorithm Delaying Health Care for Black People'; Lesley Inker, Nwamaka Eneanya, Margret Andresdottir, Josef Coresh, Hocine Tighiouart, Dan Wang, Yingying Sang, Deidra Crews, Alessandro Doria, Michelle Estrella, Marc Froissart, Morgan Grams, Tom Greene, Anders Grubb, Vilmundur Gudnason, Orlando Gutiérrez, Roberto Kalil, Amy Karger, Michael Mauer, Gerjan Navis, Robert Nelson, Emilio Poggio, Roger Rodby, Peter Rossing, Andrew Rule, Elizabeth Selvin, Jesse Seegmiller, Michael Shlipak, Vicente Torres, Wei Yang, Shoshana Ballew, Sara Couture, Neil Powe, Andrew Levey, Hrefna Gudmundsdottir, Olafur Indridason, Runolfur Palsson, Bertram Kasiske, Matthew Weir, Todd Pesavento, Harold Feldman, Amanda Anderson, Alan Go, Chi-Yuan Hsu, Arlene

Chapman, Douglas Landsittel, Michael Mrug, Alan Yu, Michael Steffes, and Barbara Braffett, 'New Creatinine and Cystatin C – Based Equations to Estimate GFR without Race.' *New England Journal of Medicine* 385 (November 2021): 1737–49; James Diao, Gloria Wu, Herman Taylor, John Tucker, Neil Powe, Isaac Kohane, and Arjun Manrai, 'Clinical Implications of Removing Race from Estimates of Kidney Function, Editorial.' *Journal of American Medical Association* 325, no. 2 (12 January 2021): 184–6.

65 Himmelfarb et al., 'The Current and Future Landscape of Dialysis.'

66 Richard Johnson, Catharina Wesseling, and Lee Newman, 'Chronic Kidney Disease of Unknown Cause in Agricultural Communities.' *New England Journal of Medicine* 380, no. 19 (May 2019): 1843–52.

67 Johnson, Wesseling, and Newman, 'Chronic Kidney Disease of Unknown Cause.'

68 Johnson, Wesseling, and Newman, 'Chronic Kidney Disease of Unknown Cause.'

69 Johnson, Wesseling, and Newman, 'Chronic Kidney Disease of Unknown Cause'; Eric Hansson, Ali Mansourian, Mahdi Farnaghi, Max Petzold, and Kristina Jakobsson, 'An Ecological Study of Chronic Kidney Disease in Five Mesoamerican Countries: Associations with Crop and Heat.' *BMC Public Health* 21 (2021): 840.

70 Himmelfarb et al., 'The Current and Future Landscape of Dialysis'; Tonelli, Vanholder, and Himmelfarb, 'Health Policy for Dialysis Care in Canada and the United States.'

71 Jeremy Greene, *The Doctor Who Wasn't There: Technology, History, and the Limits of Telehealth* (Chicago, IL: University of Chicago Press, 2022).

72 Jeremy Greene, 'As Telemedicine Surges, Will Community Health Suffer?' *Boston Review*, April 2020; Andrew Simpson, *Medical Metropolis: Health Care and Economic Transformation in Pittsburgh and Houston* (Philadelphia, PA: University of Pennsylvania Press, 2019).

Bibliography

Abraham, Laurie Kaye. *Mama Might Be Better Off Dead: The Failure of Health Care in Urban America*. Chicago, IL: University of Chicago Press, 1993.

Adas, Michael. *Dominance by Design: Technological Imperatives and America's Civilizing Mission*. Cambridge, MA: Harvard University Press, 2006.

Aronowitz, Robert. *Making Sense of Illness: Science, Society, and Disease*. Cambridge: Cambridge University Press, 1998.

Benjamin, Ruha. *Race after Technology: Abolitionist Tools for the New Jim Code*. Medford, OR: Policy Press, 2019.

Bliss, Michael. *The Discovery of Insulin*. Chicago, IL: University of Chicago Press, 1982.

Braun, Lundy. *Breathing Race into the Machine: The Surprising Career of the Spirometer from Plantation to Genetics*. Minneapolis, MN: University of Minnesota Press, 2014.

Cohen, Alan B., David C. Colby, Keith A. Wailoo, and Julian E. Zelizer (eds). *Medicare and Medicaid at 50: America's Entitlement Programs in the Age of Affordable Care*. New York: Oxford University Press, 2015.

Diao, James, Gloria Wu, Herman Taylor, John Tucker, Neil Powe, Isaac Kohane, and Arjun Manrai. 'Clinical Implications of Removing Race from Estimates of Kidney Function, Editorial.' *Journal of American Medical Association* 325, no. 2 (12 January 2021): 184–6.

Fields, Karen, and Barbara Fields. *Racecraft: The South of Inequality in American Life*. New York: Verso Books, 2012.

Fields, Robin. 'In Dialysis: Life Saving Care at Great Risk and Cost.' *ProPublica*, 9 November 2010.

Fields, Robin. 'God Help You. You're on Dialysis.' *The Atlantic*, December 2010.

Foucault, Michel. *The Birth of Biopolitics: Lectures at the College de France, 1978–1979*. New York: Picador Press, 2010.

Fox, Renee, and Judith Swazey. *The Courage to Fail: A Social View of Organ Transplants and Dialysis*. Chicago, IL: University of Chicago Press, 1978.

Gander, Jennifer, Xingyu Zhang, Katherine Ross, Adam Wilk, Laura McPherson, Teri Browne, Stephen Pastan, Elizabeth Walker, Zhensheng Wang, and Rachel Patzer. 'Association between Dialysis Facility Ownership and Access to Kidney Transplantation.' *JAMA Original Investigation* 322, no. 10 (2019): 957–73.

Greene, Jeremy. *The Doctor Who Wasn't There: Technology, History, and the Limits of Telehealth*. Chicago, IL: University of Chicago Press, 2022.

Greene, Jeremy. 'As Telemedicine Surges, Will Community Health Suffer?' *Boston Review*, April 2020.

Gutmann, Amy, and Jonathan Moreno. *Everybody Wants to Go to Heaven but Nobody Wants to Die: Bioethics and the Transformation of Health Care in America*. New York: Liveright Publishing, 2019.

Hansson, Eric, Ali Mansourian, Mahdi Farnaghi, Max Petzold, and Kristina Jakobsson. 'An Ecological Study of Chronic Kidney Disease in Five Mesoamerican Countries: Associations with Crop and Heat.' *BMC Public Health* 21 (2021): 840.

Hatch, Ryan. *Blood Sugar: Racial Pharmacology and Food Justice in Black America*. Minneapolis, MN: University of Minnesota Press, 2016.

Headrick, Daniel. *The Tools of Empire: Technology and European Imperialism in the Nineteenth Century*. New York: Oxford University Press, 1981.

Himmelfarb, Jonathan, Raymond Vanholder, Rajnish Mehrotra, and Marcello Tonelli. 'The Current and Future Landscape of Dialysis.' *Nature Reviews Nephrology* 16 (2020): 573–85.

Hsu, Caroline, and Daniel Weiner. 'Covid-19 in Dialysis Patients: Outlasting and Outsmarting a Pandemic.' *Kidney International* 98, no. 6 (December 2020): 1402–4.

Inker, Lesley, Nwamaka Eneanya, Margret Andresdottir, Josef Coresh, Hocine Tighiouart, Dan Wang, Yingying Sang, Deidra Crews, Alessandro Doria, Michelle Estrella, Marc Froissart, Morgan Grams, Tom Greene, Anders Grubb, Vilmundur Gudnason, Orlando Gutiérrez, Roberto Kalil, Amy Karger, Michael Mauer, Gerjan Navis, Robert Nelson, Emilio Poggio, Roger Rodby, Peter Rossing, Andrew Rule, Elizabeth Selvin, Jesse Seegmiller, Michael Shlipak, Vicente Torres, Wei Yang, Shoshana Ballew, Sara Couture, Neil Powe, Andrew Levey, Hrefna Gudmundsdottir, Olafur Indridason, Runolfur Palsson, Bertram Kasiske, Matthew Weir, Todd Pesavento, Harold Feldman, Amanda Anderson, Alan Go, Chi-Yuan Hsu, Arlene Chapman, Douglas Landsittel, Michael Mrug, Alan Yu, Michael Steffes, and Barbara Braffett. 'New Creatinine and Cystatin C – Based Equations to Estimate GFR without Race.' *New England Journal of Medicine* 385 (November 2021): 1737–49.

Interview with Belding Scribner, 26 March 1986. Belding H. Scribner Papers, Acc 3301–009, Tapes 1–4. University of Washington Special Collections.

Johnson, Richard, Catharina Wesseling, and Lee Newman. 'Chronic Kidney Disease of Unknown Cause in Agricultural Communities.' *New England Journal of Medicine* 380, no. 19 (May 2019): 1843–52.

Kahn, Jonathan. *Race in a Bottle: The Story of BiDiL and Racialized Medicine in a Post-Genomic Age*. New York: Columbia University Press, 2012.

Lackland, Daniel, and Michael Moore. 'Hypertension-Related Mortality and Morbidity in the Southeast.' *Southern Medical Journal* 90 (1997): 191–8.

Lanska, Douglas, and Lewis Kuller. 'The Geography of Stroke Mortality in the United States and the Concept of a Stroke Belt.' *Stroke* 26 (1995): 1145–9.

Linker, Beth. 'On the Borderland of Medical and Disability History: A Survey of the Fields.' *Bulletin of the History of Medicine* 87, no. 4 (2013): 499–535.

Madhusoodanan, Jyoti. 'Is a Biased Algorithm Delaying Health Care for Black People.' *Nature* 588 (December 2020): 546–7.

Mbembe, Achille. *Necropolitics*. Durham, NC: Duke University Press, 2019.

Mizelle, Jr, Richard M. 'Diabetes, Race, and Amputations.' *The Lancet: Art of Medicine* 397, no. 10281 (2021): 1256–7.

Mizelle, Jr, Richard M. 'Getting into Good Medical Trouble.' *Los Angeles Review of Books*, 6 November 2020.

National Kidney Foundation. 'Kidney Disease: The Basics.' www.kidney. org/news/newsroom/factsheets/KidneyDiseaseBasics.

Peitzman, Steven. *Dropsy, Dialysis, Transplant: A Short History of Failing Kidneys.* Baltimore, MD: Johns Hopkins University Press, 2007.

Pendras, Jerry. 'Experience with Patient Selection for Chronic Hemodialysis.' Unpublished manuscript from the Seattle Artificial Kidney Center, Belding Scribner Papers, ACC 3301–82–14, Restricted Files, Box 3, University of Washington Libraries Special Collections.

Pollock, Anne. *Medicating Race: Heart Disease and Durable Preoccupations Difference.* Durham, NC: Duke University Press, 2012.

Presser, Liz. 'Tethered to the Machine.' *ProPublica*, 15 December 2020.

Rettig, Richard, 'Origins of the Medicare Kidney Disease Entitlement: The Social Security Amendments of 1972,' in *Biomedical Politics*, edited by Kathi Hanna, Division of Health Sciences Policy, Committee to Study Biomedical Decision Making, and the Institute of Medicine, pp. 176–208. Washington, DC: National Academies Press, 1991.

Reverby, Susan. *Examining Tuskegee: The Infamous Syphilis Study and Its Legacy.* Chapel Hill, NC: University of North Carolina Press, 2013.

Roberts, Dorothy. 'What's Wrong with Race-Based Medicine? Genes, Drugs, and Health Disparities.' *Minnesota Journal of Law, Science & Technology* 12, no. 1 (Winter 2011): 1–21.

Rothman, David. *Strangers at the Bedside: A History of How Law and Bioethics Transformed Medical Decision-Making.* New York: Basic Books, 1992.

Scribner, Belding, and Albert Babb. 'Chronic Hemodialysis in Seattle: 1960–1966.' Unpublished manuscript. ACC 3301–05, Box 2, Belding Scribner Papers, University of Washington Libraries Special Collections.

Simpson, Andrew. *Medical Metropolis: Health Care and Economic Transformation in Pittsburgh and Houston.* Philadelphia, PA: University of Pennsylvania Press, 2019.

Tonelli, Marcello, Raymond Vanholder, and Jonathan Himmelfarb. 'Health Policy for Dialysis Care in Canada and the United States.' *Clinical Journal of the American Society of Nephrology* 15 (November 2020): 1669–80.

Tuchman, Arleen. *Diabetes: A History of Race and Disease.* New Haven, CT: Yale University Press, 2020.

Vyan, Darshali, Leo Eisenstein, and David Jones. 'Hidden in Plain Sight – Reconsidering the Use of Race Correction in Clinical Algorithms.' *The New England Journal of Medicine* 383, no. 9 (August 2020): 874–82.

Wailoo, Keith. *Drawing Blood: Technology and Disease Identity in Twentieth-Century America.* Baltimore, MD: Johns Hopkins University Press, 1999.

Wailoo, Keith. *Dying in the City of the Blues: Sickle Cell Anemia and the Politics of Race and Health.* Chapel Hill, NC: University of North Carolina Press, 2001.

Wailoo, Keith. *How Cancer Crossed the Color Line*. New York: Oxford University Press, 2011.

Weisz, George. *Chronic Disease in the Twentieth Century: A History*. Baltimore, MD: Johns Hopkins University Press, 2014.

Winner, Langdon. *The Whale and the Reactor: A Search for Limits in an Age of High Technology*. Chicago, IL: University of Chicago Press, 1989.

3

Patients, 'consumer sovereignty', and technological change: The adoption of minimally invasive surgery

Cynthia L. Tang

Throughout most of the twentieth century, the best option for many patients suffering from gallstones was to have their gallbladders removed through a three- to six-inch abdominal incision. The procedure, known as a cholecystectomy, could require patients to spend up to seven weeks in recovery with up to one week in hospital. Beginning in 1989, however, patients could have their gallbladders removed through three or four small incisions and be back to their regular activities one to seven days later. Due to the non-fatal nature of gallstones, most patients had the flexibility to decide how and when to treat their condition. As one patient in New Jersey – who had resisted conventional surgery for eight years – explained, 'The surgery didn't look very pleasant ... And having two little kids, I didn't think I had six weeks to spare.' When the less invasive procedure – or laparoscopic cholecystectomy became available in her area in May 1990, she was able to 'squeez[e] it between work [on] Monday and a parent–teacher conference on Thursday'.[1]

The high level of interest in laparoscopic cholecystectomy among patients and surgeons led one healthcare market analyst to predict that it would be used 'for a quarter of gallbladder removals by 1992 and 75 per cent by 1994'.[2] Far exceeding this projection, an estimated 80 per cent of total gallbladder removals were already being performed laparoscopically by the end of 1992.[3] This was an extraordinary rate of adoption since most general surgeons had little or no experience with laparoscopy and many required substantial training. In contrast to conventional surgery where surgeons had open access to the abdominal cavity and could directly see, feel, and manipulate the organs and tissues, laparoscopic surgery was performed from outside the body using long surgical instruments that were inserted into the abdomen through half-inch incisions. With this type of procedure, surgeons' only view into the body was through the laparoscope: a telescopic instrument (known more generally as an endoscope) that was specifically used for visualizing the abdominal cavity and which, beginning in the late 1980s, could be equipped with a camera that transmitted real-time video footage to a television monitor. Although laparoscopy had been used in gynaecological procedures, first as a diagnostic tool since the 1930s and then as an operative tool in the 1960s, the technology was largely ignored in general surgery until its application to cholecystectomies.[4] By using laparoscopic technology, surgeons' vision became mediated through video on a screen while their touch became mediated through instruments that were manipulated extracorporeally. This, however, required a wholly different skill set to operate than conventional instruments.[5] Some surgeons quickly cautioned their colleagues against being too hasty in offering the new laparoscopic gallbladder procedure, warning that '[u]nleashing it without adequate safeguards could turn out to be a surgical nightmare'.[6] Surgical organizations in the United States, however, had little power to monitor whether individual surgeons underwent sufficient training in laparoscopy and laparoscopic surgical techniques. Indeed, the early use of the technique by some surgeons led to higher complication rates ranging from relatively moderate injuries to severe organ damage, and sometimes even death.

Existing explanations for the speed with which general surgeons began to offer the laparoscopic procedure identify patient demand for less invasive gallstone treatment as a major influence.[7] Occupational sociologist James Zetka has even implied that general surgeons applied laparoscopic technology to gallbladder surgery in order to compete with the new non-surgical treatments that gastroenterologists were beginning to offer and publicize in the 1980s.[8] According to Zetka, the competition for gallstone patients prompted surgeons to quickly embrace the radically different approach as a 'response to the very real erosion of their market turf'.[9] Likewise, a 1993 National Institutes of Health report credited market forces as the predominant driver behind the rapid growth in the technique's usage.[10] Surgeons' accounts of the period similarly describe it as a 'patient-led revolution in healthcare' and highlight the overwhelming pressure many felt to offer their patients the laparoscopic procedure.[11] These narratives emphasize the belief that patients held significant power as medical consumers to influence medical practice.

The idea that patient demand can drive change in medical practice is the basis for market-oriented strategies in healthcare policy making. As historian Nancy Tomes has discussed, advocates of consumer-driven healthcare policy believe 'that simply providing patients more choice will bring about structural changes favorable to their economic and therapeutic interests'.[12] This premise is more broadly referred to by economists as 'consumer sovereignty', which posits that consumers are the best judge of their own needs, and when allowed to choose freely in a competitive marketplace, they have the power to influence the availability, cost, and quality of goods and services. Though it was originally coined during the interwar period to describe an ideal against which economic systems could be assessed, it has become a widely assumed aspect of laissez-faire, free-market economic theory.[13] But as Tomes points out, the mechanics of supply and demand in healthcare do not conform to the free-market economic model where competition between providers leads to lower prices and better quality.[14] Whether consumer sovereignty can even exist in the healthcare marketplace is subject to debate.[15] In addition to having to rely on their physicians' judgement with respect to treatment options, the degree to which patients can have consumer sovereignty often depends on the urgency of their medical treatment. In contrast to patients suffering from more acute conditions, gallstone patients have the luxury of significant consumer sovereignty and can thus impact the gallstone treatment marketplace.

This chapter explores the ways in which patient demand influenced, or has been thought to influence, the use of laparoscopic technology in gallbladder surgery. I first show that in contrast to suggestions that general surgeons developed and adopted laparoscopic cholecystectomy as a way to compete with gastroenterologists' new non-surgical treatments, the surgeons who developed the procedure had multiple motivations that extended beyond a specific demand from patients for a less invasive treatment for gallstones. Instead, the developers of laparoscopic cholecystectomy were interested in reducing the trauma of abdominal surgery and in applying laparoscopic and laser technology to general surgery.

Secondly, I demonstrate that the procedure's early adopters helped create demand by generating publicity for themselves as local pioneers through newspaper reports that announced the availability of the new procedure. Such news reporting in which surgeons took an active role in publicizing themselves and the treatments they could offer was only just becoming normalized throughout the 1980s.

Once laparoscopic cholecystectomy was introduced into a community, gallbladder patients had the necessary information and access, and thus consumer sovereignty, to place pressure on other surgeons in the region to offer the technique as well. While historians have shown how patients have successfully advocated for changes to medical practice, in the case of laparoscopic cholecystectomy, demand was generated through mass media publicization and advertising.[16] Characterizing the adoption of laparoscopic cholecystectomy as a patient demand-driven revolution disregards how early providers instigated local competition for gallbladder patients through their self-promotion. Similarly, the claim that the rapid growth in the technique's use was due to 'market forces generated, not inappropriately, by patient demand' overlooks the responsibilities that individual surgeons and hospital credentialling committees had to make unbiased judgements as to whether a procedure could be safely performed.[17] I suggest that the invocation of 'patient demand' to account for the uncontrolled adoption of laparoscopic cholecystectomy was more of a justification for the lack of restraint demonstrated by some surgeons than an accurate explanation.

Developing a less invasive gallbladder surgery

Open gallbladder surgery has a minimal risk of complications and usually resolves the problem completely by preventing the further formation of gallstones that necessitate the procedure in the first place. But by 1989, when general surgeons began to adopt laparoscopic cholecystectomy, patients could also choose to treat their gallstones using the non-surgical interventions offered by gastroenterologists. Oral bile acids, for example, could be taken in an attempt to dissolve stones – a therapy that was later enhanced with the use of ultrasonic lithotripsy to fragment the stones for better surface area contact with the solvent. Gastroenterologists could also use flexible endoscopes to access the common bile duct through the oesophagus to remove stones that were blocking the biliary system in a procedure called endoscopic sphincterotomy. Any stones that were further back in the cystic duct or still in the gallbladder would not have been accessible with this technique in the 1980s.[18] In contrast to gallbladder removal, neither of these treatments were permanent solutions and allowed the possibility for symptoms to recur.

Depending on the severity of any symptoms or complications (as well as availability in their area), patients could often take their time in considering these options before deciding on a treatment plan. According to Zetka, gallstone patients were arbiters in a turf war between general surgeons and gastroenterologists over the treatment of the gastrointestinal tract.[19] He argues, for example, that gastro-enterologists' use of lithotripsy to break up gallstones and endoscopic sphincterotomy to remove them was 'especially threatening to surgeons' because of their superior treatment outcomes.[20] In this narrative, laparoscopic cholecystectomy is implied to have been consciously developed in an effort to keep gallstone treatment within their jurisdiction.[21] Published retrospectives and oral history interviews with the developers of the technique, however, show that its development was not a coordinated attempt to reassert general surgery's claim on the treatment of gallstones. Instead, the developers of the technique each had differing motivations, including the reduction of trauma during abdominal surgery and an interest in incorporating laparoscopic and laser technology into general surgery.

The first laparoscopic cholecystectomy was performed at a small independent surgical clinic in Lyon, France, by Philippe Mouret in 1987. According to Mouret, 'the first cholecystectomy was performed quite naturally, without premeditation'.[22] For him, it was a logical progression of his work that had taken place over decades. As a surgical intern in the 1960s, Mouret trained in vascular surgery, orthopaedics, pulmonary surgery, and, significantly, gynaecology, where laparoscopic techniques more generally had their origins. After a year of military service, he returned to Lyon and eventually became *chirurgien de garde* of the urology department at L'Hôpital Edouard Herriot.[23] For Mouret, this was the decisive moment of his career. In an interview a few months before his death in 2008, Mouret explained:

I found myself facing many emergencies with patients suffering from acute stomach aches. At the time, there were no CT scans, no precise diagnostic tests ... So we had to open the patient in order to make a diagnosis. As a result, some patients were treated for nothing! With all the risks of mortality that this implied. And this, this scandalized me.[24]

Mouret's experiences at L'Hôpital Edouard Herriot inspired him to incorporate the skills that he learned in his gynaecology training into his surgical practice.[25] He felt that he could decrease the incidence of post-surgical scarring in his patients and the potential need for more operations in the future by using laparoscopy as an exploratory tool in lieu of an immediate laparotomy. As his skill in laparoscopy developed, he continued to push the boundaries of how much of a surgical procedure could be performed before an open incision was necessary.[26] In March 1987 Mouret found himself in a position where he could do so in a cholecystectomy. With each step of the procedure, it would occur to him that he could continue to delay the moment that he had to make a larger incision.[27] Going past the exploratory stage, he was able to dissect the gallbladder from its surroundings laparoscopically, then pull it out of the abdomen and excise it extracorporeally. When he checked on his patient the next morning, she was up, ready to leave, and upset at him because without the expected scar, she believed that he had not removed her gallbladder as he had promised.[28] Although Mouret himself saw the procedure he performed that day as being a laparoscopy-aided cholecystectomy, the surgical world came to see it as the first laparoscopic cholecystectomy.

For the American developers of laparoscopic cholecystectomy, it was the laparoscopic and laser technologies that provided inspiration. In the United States, the idea to develop laparoscopic cholecystectomy originated when two physicians from Marietta, Georgia, Barry McKernan and William Saye, observed a colleague remove gynaecological scar tissue using a laser inserted through a laparoscope in April 1988.[29] As a general surgeon, McKernan's reaction was that a gallbladder could also be removed with a similar technique.[30] Together, McKernan and Saye attended a course on laser surgery where they met Eddie Reddick, a surgeon from Nashville, Tennessee, who was well known for his work using lasers in the treatment of haemorrhoids and skin lesions.[31] Reddick frequently taught courses on the surgical applications of laser technology as a spokesperson for laser instrument companies.[32] After speaking to McKernan and Saye, Reddick also became interested in developing a laser laparoscopic cholecystectomy and began to strategize about how to do so with his partner, Douglas Olsen.[33] Since at the time lasers were more commonly used in gynaecological procedures and had yet to be widely applied to general surgery, much of their efforts were in the interest of assisting laser manufacturers to expand into the general surgery market.[34] Reddick's first attempt to complete a cholecystectomy laparoscopically had to be converted to open surgery because he was unable to get control of the cystic duct.[35] To solve the problem, Olsen modified a US Surgical M11 clip applier that was originally designed for use in open surgery so that it could be used through the laparoscopic cannulas. The Nashville team were able to complete their first laparoscopic cholecystectomy using Olsen's 'jerry-rigged' instrument in September 1988, three months after McKernan and Saye performed their first procedure in Marietta.[36]

Whereas Mouret's narrative primarily focuses on his motivation to minimize abdominal trauma, the American surgeons were inspired by the laparoscopic and laser technologies and their interests in applying them to procedures in general surgery. Notably, these explanations do not include a concern that general surgeons were losing patients to gastroenterologists' non-surgical gallstone treatments. Though other historical studies have demonstrated the patient's role in shaping medical practice through advocacy and the accumulation of expertise, in the case of laparoscopic cholecystectomy, demand for less invasive gallbladder surgery was rapidly generated only after it was developed and publicized.

Generating demand for laparoscopic cholecystectomy

There was indeed significant demand for the minimally invasive procedure once patients learned of its availability. The earliest laparoscopic cholecystectomy patients learned that such a procedure was possible simply by being in contact with the right people at the right time. A patient named Julie Musselman, for example, met William Saye during a chance encounter at a barbershop in June 1988, two months after he and Barry McKernan began considering the idea of a laparoscopic gallbladder removal.[37] Musselman had been postponing gallbladder surgery because 'she was living in Florida, [and] was concerned about a post-operative scar'.[38] According to Saye, after learning that it would soon be possible to remove gallbladders through three or four small punctures, she enthusiastically replied, 'I want to have it now!'[39] Musselman became the first laparoscopic cholecystectomy patient in the United States on 22 June 1988.[40] Another patient, Barbara Kornblau, learned about the new procedure because her obstetrician had a colleague who was pursuing training in laparoscopic gallbladder surgery.[41] Three weeks after giving birth, she became the first laparoscopic cholecystectomy patient in Miami.[42] Similarly, Lisa Tharp, the first laparoscopic cholecystectomy patient in Virginia, 'had been prepared to undergo traditional gall bladder surgery' but happened to consult her surgeon, Dr William E. Kelley, just days before he left for a training course in the new technique.[43] As the *Times-Dispatch* of Richmond, Virginia, reported, '[Kelley] thought it only fair he mention [the] possible alternative to his gall bladder patients.'[44]

Musselman, Kornblau, and Tharp were among the many laparoscopic cholecystectomy patients to be featured in news coverage publicizing the procedure in the early 1990s. Beginning in August 1989, local newspapers throughout the United States began reporting on the availability of a new and less invasive treatment for gallstones.[45] In addition to informing the public about how the new procedure was less painful, less expensive, and required less recovery time,[46] articles often commemorated the first laparoscopic cholecystectomy performed in a city or state and celebrated the first surgeons or hospitals to provide the new procedure as local pioneers.[47] These stories were frequently reprinted or reported on separately in the newspapers of the surrounding area. The local Richmond newspaper, for instance, reported Lisa Tharp's gallbladder surgery just two days after her procedure. At least one other publication in the surrounding area picked up the story shortly after.[48] The *Miami Herald* included Barbara Kornblau's experience as part of a feature piece on the new 'Nintendo Surgery' for gallbladder removal, describing it as 'the hottest thing to happen in surgery in years'.[49] The article was later reprinted throughout the country, for example, in the *Greenville News* of Greenville, North Carolina, the *Fresno Bee* of

Fresno, California, the *Hartford Courant* of Hartford, Connecticut, and the *Morning Call* of Allentown, Pennsylvania.[50]

In itself, the coverage of medical breakthroughs in popular news media was not unusual.[51] However, the way in which individual surgeons and their hospitals solicited patients for laparoscopic cholecystectomy through media interviews and advertising would have been considered unethical just ten years earlier.[52] The practice was normalized over the course of the 1980s after a 1982 Supreme Court ruling that compelled the American Medical Association to remove restrictions on physician advertising from their ethical code. By the time surgeons began offering laparoscopic gallbladder removal in 1989, the stage was set for the procedure to be extensively promoted to the public. Unsurprisingly, many of the local news reports announcing its availability were the result of public relations efforts made by surgeons and hospitals. As one journalist explained, after attending laparoscopic cholecystectomy training courses, surgeons would often 'go home, call their hospital public relations offices to spread the word, and start working on real patients'.[53] Arnot-Ogden Memorial Hospital in Elmira, New York, for example, held a press conference for one of their surgeons when he returned from training to announce plans to introduce the procedure to the community a few weeks later.[54] Other press conferences occurred after surgeons performed their first laparoscopic cholecystectomy cases and included the recovering patients.[55] According to Barbara Kornblau, her surgeon also invited her to participate in local radio shows and a nationally broadcast television interview, in addition to her interview with the *Miami Herald*.[56]

While some news reports remained stand-alone stories in local newspapers, others were part of larger promotional campaigns. Michael Gleeson's first three procedures at Carbondale General Hospital on 10 May 1990 were reported the next day in the *Scranton Times* and in Scranton's *Tribune* as '[h]istory of sorts [being] made'.[57] Following the news article were a series of advertisements promoting that Dr Gleeson and Carbondale General Hospital were the first to offer 'laparoscopic laser gallbladder surgery' in the northeastern and central Pennsylvania and mid and southern New York regions.[58] The advertisement appeared in newspapers throughout Pennsylvania, New York State, and New Jersey.[59] Later advertisements relayed 'CONGRATULATIONS to

Michael Gleeson, M.D. on successfully performing his 200th Laser Laparoscopic Bellybutton Gallbladder Removal' and promoted him as the most experienced laparoscopic cholecystectomy provider in the area.[60]

The extensive publicization of laparoscopic cholecystectomy played a key role in generating interest for the procedure. Mary Theresa Garbera, the first patient to undergo the procedure at Carbondale General Hospital, originally heard about the technique from a television programme: 'They were doing the bellybutton surgery in California, so I questioned my doctor about it ... I definitely didn't want traditional surgery.'[61] As one of Gleeson's patients, her surgery was publicized the day after it took place and her experience was included in an article in the *Scranton Times* two months later.[62] The newspaper coverage of Garbera's surgery in turn prompted Scranton resident Marianne Popish to seek out the procedure for herself.[63] Since the first physician she consulted told her that gallbladder surgery would require a week of hospitalization and an additional six weeks of recuperation, Popish 'urged anyone with gallbladder problems to get a second opinion if a doctor recommends the traditional surgery'.[64] As she told Scranton's *Tribune*, 'I just can't believe that people chose the other option knowing this is available. I feel they're being misled.'[65] Publicity for the procedure at the Methodist Hospital of Hattiesburg, Mississippi, had a similar effect. Just three weeks after John Bagnato's first laparoscopic cholecystectomy made headlines in the *Hattiesburg American*, the *Clarksdale Press Register*, and the *Greenwood Commonwealth*, he had 'patients coming to Hattiesburg from all over the state ... People are staying in motel rooms until their surgeries. It's really generating business'.[66] Such publicity was clearly effective in creating demand for laparoscopic cholecystectomy among patients suffering from gallstones.

Gallstones and consumer sovereignty

The widespread reporting on laparoscopic cholecystectomy provided gallstone patients with the necessary knowledge to make decisions about their treatment plans without having to rely solely on information from the surgeons they consulted. Since most gallstone cases do not require immediate attention and can often be managed through dietary changes, many patients could choose to postpone treatment until the procedure was available in their area. As one patient in San Antonio, Texas, told a reporter in January 1990, though she 'was suffering stabs of pain in her right side ... she refused to undergo the conventional technique and insisted on waiting until preparations were finished for the new form of the procedure to be offered'.[67] Similarly, a patient in Olyphant, Pennsylvania, told a reporter that she had originally planned to have open gallbladder surgery 'but waited as long as I could until this new procedure was available'.[68] Some patients even cancelled their already scheduled cholecystectomies after learning about the less invasive technique in order to reschedule with a surgeon who was offering the laparoscopic procedure. Commending *The News Journal* of Wilmington, Delaware, for its front page reporting about the availability of laparoscopic gallbladder surgery at Nanticoke Hospital, one reader wrote: 'My wife was scheduled for the standard method of gallbladder surgery on May 21. After seeing the story we called Dr Frederick Toy ... and my wife's surgery was re-scheduled for May 24.'[69] That gallstone patients had such flexibility in deciding when to undergo treatment afforded them significant consumer sovereignty as compared to patients who required more urgent care.

Cholecystectomy patients also constituted a large enough consumer group that surgeons felt pressured into adopting the laparoscopic procedure. Unlike patients in need of a heart transplant, for instance, gallstones are a common enough ailment that gallbladder surgery has been described as one of general surgeons' 'bread-and-butter' procedures.[70] A commonly told cautionary tale of the period describes the surgeon who refused to learn how to perform a laparoscopic cholecystectomy and lost his practice because patients were no longer willing to undergo the open procedure.[71] As one surgeon explained in a 2014 interview, '[C]holecystectomy was such a common operation and such an important part of most [general] surgeons' practice, that if you didn't do cholecystectomy by laparoscopy, there was a great threat to your practice, and you may lose lots and lots of patients.'[72] Edward Teitel, a surgeon in Ozark, Alabama, for example, waited for six months to attend one of Reddick's courses and an additional two months to receive the laparoscopic instruments.[73] By the time he performed his first laparoscopic cholecystectomy in September 1990, 'he had already lost patients to the hospital in nearby Dothan'.[74]

Together with the extensive publicization of laparoscopic chole-cystectomy, the common and non-fatal nature of gallstones allowed patients to have sufficient consumer sovereignty and power as a consumer group to pressure surgeons and hospitals to expand its availability as quickly as possible. This is particularly noteworthy considering that it took some effort to convince insurance providers to cover the laparoscopic option. As Douglas Olsen recalls, 'One of the things we had to struggle with in the early days was getting reimbursed because a lot of the insurance companies would take the posture that this is an experimental [procedure].'[75] He was able to persuade one insurance executive to cover laparoscopic cholecys-tectomy by agreeing to keep patients overnight instead of providing it as an outpatient surgery.[76] Patients also played a role in advo-cating for coverage, with many contacting their insurance provid-ers to argue that their preferred treatment would allow them to return to work earlier in addition to reducing their hospital bills.[77] Though the national Blue Cross and Blue Shield Association did not have a collective recommendation, roughly half of the nation's seventy-four independent plans extended coverage for laparoscopic gallbladder removal by August 1990.[78] But regardless of insurance coverage, it is clear that enough patients refused open surgery after learning about laparoscopic cholecystectomy that many surgeons felt compelled to seek laparoscopic training as quickly as possible.

Consequences of competition in the surgical marketplace

Beyond the prestige that came with being the first in the area to perform laparoscopic cholecystectomy, there were major economic incentives to being the only surgeon or hospital offering it. As it was later explained in the *Wall Street Journal*, 'The first one in the community who does it gets all the gallbladders for a while.'[79] The economic importance of cholecystectomy to a general surgeon's practice drove some surgeons to hastily offer it before getting ade-quate training, in order to compete for patients. As the *Philadelphia Daily News* reported in June 1990, 'There is a lot of competition among local surgeons to learn the technique and begin practicing this procedure.'[80] With 'hundreds of surgeons … lining up for courses to learn the technique', the high demand for training in laparoscopic cholecystectomy quickly created a seller's market in laparoscopic surgery training courses.[81] Just weeks after performing his first laparoscopic cholecystectomy in December 1989, John Bagnato was already 'hop[ing] to offer classes to teach other surgeons the techniques by March [1990]'. By August, it was reported that he had provided training for one hundred surgeons throughout the United States, including every surgeon in his town.[82] Similarly, surgeons who were the first to perform the procedure in the state of Wisconsin in December 1989 announced that they would be running a two-day training programme at the beginning of April 1990.[83] According to the *Wall Street Journal*, Eddie Reddick and William Saye also trained more than 1,300 doctors between June and December 1990 at a training centre financed by a laparoscopic instrument manufacturer.[84]

Already by mid-1990, however, there were concerns that too many surgeons were being trained too fast in courses that varied in length and quality. As the *New York Times* reported, 'Some courses last a day, other three days. In some courses surgeons practice on pigs, whose gallbladder anatomy most closely resembles that of humans. In other courses, no animal work is done.'[85] A study at the University of Pennsylvania's Institute of Health Economics found that among surgeons who adopted the technique in 1989 and the first quarter of 1990, 35 per cent performed their first laparoscopic cholecystectomy without having previously assisted in one with a more experienced colleague. When performing their first laparoscopic cholecystectomy, 58 per cent of surgeons did so without supervision from a more experienced surgeon.[86]

The rush to offer the procedure before acquiring adequate experience sometimes had dire consequences. As the *New York Times* later reported in 1992, 'Among the cases of botched surgery … was that of a 66-year-old woman who bled to death after a surgeon accidentally punctured her aorta … The surgeon, who had learned the laparoscopic technique in a one-day seminar, completed the operation before recognising the source of bleeding.'[87] In another case, 'a 31-year-old woman who ended up requiring multiple surgeries to reconstruct parts of the liver and bile system … was left with a permanent risk of repeated episodes of inflammation of the bile ducts that can be painful and life-threatening. Her surgeon had attended a two-day training program.'[88] In September 1992 the National Institutes of Health convened a Consensus Development Conference on Gallstones and Laparoscopic Cholecystectomy to 'focus on the evaluation of emerging data and controversies surrounding laparoscopic cholecystectomy'.[89] The resulting statement, published in the *American Journal of Surgery* in April 1993, estimated that rates of bile duct injuries could be up to one in two hundred patients who underwent the laparoscopic procedure compared to one in five hundred patients who underwent open cholecystectomy.[90] According to the Physician Insurers Association of America, this was accompanied by a significant increase in malpractice lawsuits involving gallbladder surgery: 189 claims for bile duct injury during laparoscopic cholecystectomy between 1990 and 1993, up from thirty-five in open procedures between 1985 and 1990.[91]

Though individual surgeons could refrain from operating until they received adequate training, discussions about the rapid adoption of laparoscopic cholecystectomy and its high rate of complications tended to place responsibility for controlling the use of the technique on hospitals. Indeed, surgical organizations in the United States maintained they had little power to restrict the availability of laparoscopic cholecystectomy or to monitor whether individual surgeons acquired enough training before offering the procedure. According to the *Los Angeles Times*, 'The biggest surgeons' group, the American College of Surgeons, insist[ed] laparoscopic-credentialing is up to individual hospitals.'[92] Hospitals, however, had considerable financial motivations to offer the procedure as quickly as possible. The desire to provide the most advanced medical services sometimes created a dangerous conflict of interest with regards to a hospital's ability to regulate whether a surgeon should be certified to perform a new procedure. As the *Los Angeles Times* explained, 'Since most surgeries require neither national consensus nor government approval, the [individual hospital credentialling] committees are the only check on what operations doctors can perform.'[93] Yet these committees frequently include the very surgeons under review (or their colleagues) and are often asked to make decisions that can have a significant impact on the incomes of all parties involved.

By early 1992 this lack of oversight became unacceptable to regulators at the New York State Health Department. In June 1992 it issued guidelines to hospitals throughout the state stipulating that privileges for laparoscopic cholecystectomy could only be granted to surgeons with experience in assisting an already-privileged surgeon in at least five procedures and in performing at least ten additional procedures as the responsible surgeon under the supervision of an already-privileged surgeon.[94] Unlike previous guidelines suggested by surgical organizations such as the Society of American Gastrointestinal Endoscopic Surgeons, these guidelines were more specific with regards to the amount of supervised practical experience that surgeons had to have in order to be granted privileges in laparoscopic surgery. And, importantly, these guidelines were mandated by the state government. But despite the increase in gallbladder-related complications, the adoption of laparoscopic cholecystectomy is often referred to as the beginning of the 'laparoscopic revolution' in general surgery.[95] Its development was a successful proof of concept that laparoscopic techniques had a place in general surgery and led to the expansion of their use for hernia repair, bowel resections, and bariatric surgery, to name just a few of the procedures that are now routinely performed via laparoscopy. The explanation that patient demand was the major driving force behind its uncontrolled adoption, however, overlooks critical gaps in the professional oversight of surgical innovation and could be taken to imply that potentially negligent behaviour is justifiable when commercial interests are at stake.

Concluding remarks

Examining the histories of patients seeking out specific treatments that they learn about outside the doctor's office provides an opportunity to understand the circumstances in which consumer sovereignty can exist in medical care and how it can affect medical practice. For some analysts of healthcare economics, 'consumer sovereignty is more of a fiction than fact in the healthcare market' because in order to have true consumer sovereignty, consumers must have both freedom of choice and the opportunity to judge information about quality and value.[96] Patients' accounts of refusing to undergo open gallbladder surgery and of specifically seeking surgeons who offered the new procedure demonstrate the relative freedom that they had in their medical decision making. It is unlikely, however, that patients were able to judge the quality of their surgeons' skill in laparoscopic surgery. But instead of thinking about consumer sovereignty in absolute terms, it is perhaps more useful as an ideal against which cases can be considered. The characterization of laparoscopic cholecystectomy's adoption as a 'patient-driven revolution' suggests that gallstone patients did have enough consumer sovereignty to influence the gallstone treatment marketplace.

Similarly, by publicizing that they were offering the new procedure through advertising and local news media, surgeons and hospitals were soliciting cholecystectomy patients and, in essence, treating them as consumers. The economic importance of gallbladder surgery placed pressure on the other surgeons and hospitals in the region to begin offering the laparoscopic procedure as quickly as possible. The market-driven adoption of laparoscopic cholecystectomy was thus less about demand from patients and more about competition among surgeons and hospitals. Though surgical organizations issued guidelines for training and credentialling, decisions over whether a surgeon could perform a laparoscopic cholecystectomy were ultimately at the discretion of their hospital, which had obvious conflicting interests. Since many patients opted to postpone treatment until the laparoscopic procedure was available to them, 'patient demand' may have seemed like a reasonable justification for rushing the adoption of the new technique. Still, the emphasis on patient demand as the explanation for how quickly this major technological change in surgical practice took place obscures the lack of unbiased regulation in surgical innovation. Although those seeking less invasive cholecystectomies were treated as consumers, they were still patients that needed responsible medical judgement from their surgeons.

Notes

1 Judy Holmes, 'A Cut Above: Technique Cuts Recovery Time and Risk for Gallbladder Patients.' *Asbury Park Press*, 1 May 1990, B1–2.
2 Malcolm Ritter, 'Gallbladder Removal Avoids Incision, Speeds Up Recovery, Causes Less Pain.' *Lancaster Eagle-Gazette*, 13 August 1990, p. 13.

3 National Institutes of Health, 'National Institutes of Health Consensus Development Conference Statement on Gallstones and Laparoscopic Cholecystectomy.' *American Journal of Surgery* 165 (1993): 390–6.

4 Nicholas Whitfield, 'A Revolution through the Keyhole: Technology, Innovation, and the Rise of Minimally Invasive Surgery', in *The Palgrave Handbook of the History of Surgery*, edited by Thomas Schlich, pp. 525–48 (London: Palgrave Macmillan, 2018). For the use of laparoscopy in gynaecology, see Ramona Braun, 'Laparoscopy as a Neo-Eugenic Practice in Gynaecology, 1940s–60s' (PhD dissertation, University of Cambridge, 2015). For more on general surgeons' resistance to adopting endoscopic/laparoscopic technology in surgical procedures, see Cynthia L. Tang, 'Technological Change in "Ordinary Medicine": The Emergence of Minimally Invasive Gallbladder Surgery, c. 1970–1992' (PhD dissertation, McGill University, 2021).

5 For a more in-depth discussion of how laparoscopic surgery differs from open surgery, see Rachel Prentice, 'Swimming in the Joint', in *Bodies in Formation: An Ethnography of Anatomy and Surgery Education*, pp. 171–98 (Durham, NC: Duke University Press, 2012).

6 Alfred Cuschieri, 'The Laparoscopic Revolution – Walk Carefully before We Run.' *Journal of the Royal College of Surgeons of Edinburgh* 34 (1989): 295. This warning was reiterated a few months later: Alfred Cuschieri, George Berci, and Charles K. McSherry, 'Laparoscopic Cholecystectomy.' *American Journal of Surgery* 159 (1990): 273.

7 See, for example, Herschel A. Graves, Jeanne F. Ballinger, and William J. Anderson, 'Appraisal of Laparoscopic Cholecystectomy.' *Annals of Surgery* 213 (1991): 655–62; George Berci, 'Laparoscopic Cholecystectomy Viewed from the USA.' *Australian and New Zealand Journal of Surgery* 61 (1991): 249–50; Kalser et al., 'Conference Statement', 393.

8 James Zetka, *Surgeons and the Scope* (Ithaca, NY: Cornell University Press, 2003), pp. 137–56.

9 Zetka, *Surgeons and the Scope*, p. 153.

10 In contrast to Zetka's analysis, the referenced market was specifically for gallbladder surgery and competition for patients was among general surgeons themselves. See Kalser et al., 'Conference Statement', 393.

11 For example, see Ritter, 'Gallbladder Removal', p. 13.

12 Nancy Tomes, 'Patients or Health-Care Consumers? Why the History of Contested Terms Matter', in *History and Health Policy in the United States*, edited by Rosemary Stevens, Charles E. Rosenberg, and Lawton Burns, pp. 83–110 (New Brunswick, NJ: Rutgers University Press, 2006), p. 87.

13 Maxime Desmarais-Tremblay, 'W.H. Hutt and the Conceptualisation of Consumers' Sovereignty.' *Oxford Economic Papers* 72 (2020): 1050–71.

14 Tomes, 'Patients or Health-Care Consumers?', pp. 87–8.

15 For example, see M. Joseph Sirgy, Dong-Jin Lee, and Grace B. Yu, 'Consumer Sovereignty in Healthcare: Fact or Fiction?' *Journal of Business Ethics* 101 (2011): 459–74.

16 See, for example, Barron H. Lerner, *The Breast Cancer Wars: Hope, Fear, and the Pursuit of a Cure in Twentieth-Century America* (Oxford: Oxford University Press, 2001); Julie Anderson, Francis Neary, and John Pickstone, *Surgeons, Manufacturers and Patients: A Transatlantic History of Total Hip Replacement* (Basingstoke: Palgrave Macmillan, 2007); Steven Epstein, *Impure Science: AIDS, Activism, and the Politics of Knowledge* (Berkeley, CA: University of California Press, 1996); Beth Linker, 'Prosthetic Imaginaries: Spinal Surgery and Innovation from the Patient's Perspective', in *Technological Change in Medicine: Historical Perspectives on Innovation*, edited by Thomas Schlich and Christopher Crenner, pp. 100–28 (Rochester, NY: University of Rochester Press, 2017).

17 Kalser et al., 'Conference Statement', 393.

18 Though it is now possible to retrieve stones from the cystic duct and sometimes the gallbladder, this still requires a high level of skill.

19 Zetka, *Surgeons and the Scope*, pp. 120–35.

20 Zetka, *Surgeons and the Scope*, p. 124.

21 Zetka, *Surgeons and the Scope*, p. 137.

22 Philippe Mouret, 'Special Lecture: How I Developed Laparoscopic Cholecystectomy.' *Annals of the Academy of Medicine* 25 (1996): 744–7, 746. The original lecture was delivered at the ELSA Congress in Singapore on 8 August 1993.

23 Translated from 'Philippe Mouret, l'inventeur lyonnais de la coelioscopie, est mort', *LyonMag*, 24 June 2008. www.lyonmag.com/article/8141/philippe-mouret-l-8217-inventeur-lyonnais-de-la-coelioscopie-est-mort. Accessed 2 July 2017.

24 Translated from 'Philippe Mouret, l'inventeur lyonnais'.

25 Mouret, 'How I Developed', 745.

26 Mouret, 'How I Developed', 745.

27 Mouret, 'How I Developed', 746.

28 Mouret, 'How I Developed', 746–7.

29 Barry McKernan, Interview by Cynthia L. Tang, Marietta, Georgia, 4 June 2018. This is also described in Grzegorz S. Litynski, *Highlights in the History of Laparoscopy: The Development of Laparoscopic Techniques – A Cumulative Effort of Internists, Gynecologists, and Surgeons* (Frankfurt/Main: Barbara Bernert Verlag, 1996), p. 229.

30 McKernan, Interview.
31 Litynski, *Highlights*, p. 230.
32 Douglas Olsen, Interview with Cynthia L. Tang, Nashville, Tennessee, 29 May 2018.
33 Olsen, Interview.
34 Olsen, Interview.
35 Olsen, Interview.
36 Olsen, Interview.
37 Ron Winslow, 'Cutting Edge: A Tiny TV Camera Is Fast Transforming Gallbladder Surgery.' *Wall Street Journal*, 10 December 1990, A1, A7; Grzegorz S. Litynski, 'The American Spirit Awakens', in Litynski, *Highlights*, p. 231.
38 Litynski, *Highlights*, p. 231.
39 Litynski, *Highlights*, p. 231.
40 Litynski, *Highlights*, p. 231.
41 Barbara Kornblau, Interview with Cynthia L. Tang, 11 November 2021.
42 Ena Naunton, '"Chopsticks" Turn Rough Surgery into Easy Takeout.' *Miami Herald*, 16 March 1990, pp. 1E–2E.
43 Anna Barron Billingsley, 'New-Surgery Pioneer Is Model Patient'. *Richmond Times-Dispatch*, 27 January 1990, p. 17.
44 Billingsley, 'New-Surgery Pioneer', p. 17.
45 For example, see Richard D. Walton, 'New Method Eases Pain of Gallbladder Removal.' *Indianapolis Star*, 19 November 1989, pp. B1, B6; Jonathan Bor, 'New Technique Eases Gallbladder Surgery: Surgeons at UM Use Laparoscope.' *Baltimore Sun*, 27 October 1989, pp. 1A, 8A.
46 See, for example, Shari Roan, '"Keyhole" Incisions Are Making Surgery Less Painful, Cheaper.' *Los Angeles Times*, 3 July 1990, pp. E1–2; Rebecca Perl, 'New Procedure to Remove Gallbladder Reduces Cost, Eliminates Hospital Stay.' *Atlanta Constitution*, 2 January 1990, p. C4.
47 See, for example, Steve Twedt, 'Gallbladder Technique Used Here First Time.' *Pittsburgh Press*, 20 October 1989, p. D4; 'Laser Surgery First in State.' *Greenwood Commonwealth*, 18 December 1989, p. 3; Deborah Skipper, 'The Sky's the Limit for Laser Surgery Applications: Doctors at the Surgery Clinic of Hattiesburg Were the First to Perform Laser Laparoscopic Cholecystectomy.' *Clarion-Ledger*, 18 March 1990, p. C1; 'Atlanta Physician Pioneers Laser Gallbladder Surgery.' *Atlanta Daily World*, 25 January 1990, p. 5; 'Methodist Pioneers Laser Gallbladder Removal.' *Hattiesburg American*, 25 January 1990, p. 5D; Charlene Nevada, 'Gallbladder Surgery Technique Being Done in Ohio.' *Akron Beacon-Journal*, 13 February 1990, p. B2.

48 Billingsley, 'New-Surgery Pioneer', p. 17; 'Woman First in State to Undergo Innovative Surgery', *Daily Press*, 29 January 1990, p. B4.

49 Naunton, 'Chopsticks', 1E–2E.

50 Ena Naunton, 'Doctors Call It "Nintendo Surgery".' *The Greenville News*, 22 March 1990, pp. B1, B7; Ena Naunton, 'New Type of Gallbladder Operation.' *Fresno Bee*, 27 March 1990, pp. A10–11; Ena Naunton, 'New Bellybutton Surgery Excises Sick Gallbladders.' *Hartford Courant*, 29 March 1990, p. E6; Ena Naunton, 'Now There's a Smaller Cut for Gallbladder Surgery', *Morning Call*, 3 June 1990, pp. 15–16.

51 American popular news media routinely featured new medical treatments in their coverage, a practice that began with the Pasteur's rabies vaccine in the mid-1880s: Bert Hansen, *Picturing Medical Progress from Pasteur to Polio: A History of Mass Media Images and Popular Attitudes in America* (New Brunswick, NJ: Rutgers University Press, 2009). See also Susan Lederer, *Flesh and Blood: Organ Transplantation and Blood Transfusion in Twentieth-Century America* (New York: Oxford University Press, 2008); Ayesha Nathoo, *Hearts Exposed: Transplants and the Media in 1960s Britain* (Basingstoke: Palgrave Macmillan, 2009).

52 For more on the shift in American medical practice through which direct and indirect physician advertising became normalized, see Cynthia L. Tang, 'Physician Advertising and the "Patient-Driven Revolution" in Surgery, c. 1970–1990' (forthcoming).

53 Naunton, 'Chopsticks', 1E–2E.

54 Elizabeth Sekellick, 'New Hope for Gallbladder Patients.' *Star-Gazette*, 21 August 1990, p. B1.

55 See, for example, Jenny Labalme, 'RMC Lasers Cut Surgery Time.' *Anniston Star*, 9 March 1990, pp. 13A, 15A; Barbara Mulhern, 'New Gallbladder Surgery Technique Sharply Cuts Recuperation Period.' *Capital Times*, 12 June 1990, p. 3A.

56 Kornblau, Interview.

57 'Gall Bladder Removal Surgery Involves Innovative Technique.' *Scranton Times*, 11 May 1990, p. 17; Jerry Palko, 'Surgical Technique Introduced in Region.' *Tribune*, 11 May 1990, pp. A3, A14.

58 Carbondale General Hospital, 'Which of These Women Had Gallbladder Surgery Yesterday?' *Tribune*, 8 June 1990, p. C4.

59 See, for example, *Scranton Times*, 7 June 1990, p. 18; *Press & Sun-Bulletin*, 7 June 1990), p. 5D; *Ithaca Journal*, 11 June 1990, p. 14A; *Citizens' Voice*, 16 June 1990, p. 23; *Times Leader*, 17 June 1990, p. 16C; *Pottsville Republican*, 20 June 1990, p. 34; *Press-Enterprise*, 20 June 1990, p. 27; *Morning Call*, 25 June 1990, p. C18; *Daily Record*, 31 July 1990, p. A15; *Star-Gazette*, 5 August 1990, p. 7A.

60 'CONGRATULATIONS to Michael Gleeson...' *Scranton Times*, 23 October 1990, p. 7; Carbondale General Hospital, 'Another Medical First at Carbondale General Hospital.' *Scranton Times*, 29 November 1990, p. 24.

61 'Lasers Make Surgery Easier at Carbondale General.' *Scranton Times*, 20 July 1990, p. 29.

62 'Innovative Technique', p. 17; Palko, 'Surgical Technique', pp. A3, A14; 'Lasers Make Surgery Easier', p. 29.

63 Margaret Emery, 'Simplifying Surgery.' *Tribune*, 13 August 1990, pp. C1, C5.

64 Emery, 'Simplifying', pp. C1, C5.

65 Emery, 'Simplifying', pp. C1, C5.

66 Kelly Carson, 'Doctor Zaps Gallbladder for 1st Operation in State: New Laser Surgery Saves Recovery Time.' *Hattiesburg American*, 17 December 1989, pp. 1A, 18A; 'Laser Gallbladder Surgery Cuts Time in Hospital, Recuperation.' *Clarksdale Press Register*, 18 December 1989, p. 3B; 'Laser Surgery First in State', p. 3; Kelly Carson, 'Laser Surgery May Be Taught Here.' *Hattiesburg American*, 8 January 1990, p. 5A.

67 Mark Linsalata, 'Laser Shines in Surgery on Gallbladder.' *San Antonio Express-News*, 31 January 1990, p. C1.

68 'CGH Uses Laser Surgery for Gall Bladder.' *Times Tribune*, 15 July 1990, p. A10.

69 Theodore Hooven, 'Gallbladder Article Timely.' *News Journal*, 2 July 1990, p. A9.

70 Zetka, *Surgeons and the Scope*, p. 121.

71 Jacques Vignal, Interview with Cynthia L. Tang, Lyons, France, 2 July 2018.

72 Gerald Fried, Interview by Thomas Schlich and Cynthia Tang, Montréal, Canada, 13 June 2014.

73 Winslow, 'Cutting Edge', pp. A1, A7.

74 Winslow, 'Cutting Edge', pp. A1, A7.

75 Douglas Olsen, Interview by Cynthia L. Tang, Nashville, Tennessee, 31 May 2018.

76 Insurance companies encouraged the growth of outpatient surgery in the early 1980s by covering more of the patient's bill than would be covered if they were admitted as an inpatient – for example, 100 per cent versus 80 per cent. But, as Olsen explained, inpatient surgery was billed on a per diem basis, and laparoscopic cholecystectomy patients could usually be discharged after one night rather than the average three–five nights after open cholecystectomy. Insurance companies therefore saved much more by insisting that laparoscopic cholecystectomy patients were admitted as inpatients.

77 Joyce Teveen, 'Medicine Gets Less Incisive: Doctors Using Lasers to Remove Gallbladders.' *Argus Leader*, 22 May 1990, pp. 1C, 4C.
78 Ritter, 'Gallbladder Removal', p. 13.
79 Winslow, 'Cutting Edge', pp. A1, A7. See also Bob Groves, 'New Light on Gallbladder Surgery: Doctors Embrace Laser Method.' *The Record*, 27 July 1990, pp. B5, B7.
80 Mary Flannery, 'The Changing "Scope" of Surgery: New Technique Minimizes Scarring, Allows Faster Recovery.' *Philadelphia Daily News*, 6 June 1990, pp. 35–6.
81 Winslow, 'Cutting Edge', pp. A1, A7.
82 Chantel Foretich, 'Hub Doctor Led State in New Surgery.' *Hattiesburg American*, 24 July 1990, p. 32C.
83 Karen B. Tancill, 'Surgeons Are Taking the Navel Route to Remove Gall Bladders. But It's No Gut Issue.' *Journal Times*, 14 March 1990, p. C1.
84 Winslow, 'Cutting Edge', pp. A1, A7.
85 Lawrence K. Altman, 'Complicated Surgery through Tiny Incisions.' *New York Times*, 14 August 1990, pp. C1–2.
86 José J. Escarce, Judy A. Shea, and J. Sanford Schwartz, 'How Practicing Surgeons Trained for Laparoscopic Cholecystectomy.' *Medical Care* 35 (1997): 291–6, 292.
87 Lawrence K. Altman, 'Surgical Injuries Lead To New Rule: New York Assails the Training for a Popular Technique.' *New York Times*, 14 June 1992, pp. 1, 47.
88 Altman, 'Surgical Injuries', pp. 1, 47.
89 John L. Gollan, Sarah Kalser, Henry Pitt, and Steven Strasberg, 'Foreward: Gallstones and Laparoscopic Cholecystectomy.' *American Journal of Surgery* 165 (1993): 387–8.
90 Kalser et al., 'Conference Statement', 394.
91 Kenneth A. Kern, 'Medicolegal Analysis of Bile Duct Injury during Open Cholecystectomy and Abdominal Surgery.' *American Journal of Surgery* 168 (1994): 217–22, 217.
92 Harris Meyer, 'Danger on the Cutting Edge?' *Los Angeles Times*, 29 July 1992, pp. B10–11.
93 Meyer, 'Danger', pp. B10–11.
94 Frederick L. Green, 'New York State Health Department Ruling – A "Wake-Up Call" for All.' *Surgical Endoscopy* 6 (1992): 271; New York State Department of Health, 'Laparoscopic Surgery.' New York State Department of Health Memorandum – Series 92–20, Albany, 12 June 1992.
95 See, for example, Leon Morgenstern, 'An Unsung Hero of the Laparoscopic Revolution: Eddie Joe Reddick, MD.' *Surgical*

Innovation 15 (2008): 245–8; David W. Page, *The Laparoscopic Surgery Revolution: Finding a Capable Surgeon in a Rapidly Advancing Field* (Santa Barbara, CA: Praeger, 2017).
96 Sirgy, Lee, and Yu, 'Consumer Sovereignty in Healthcare', 462.

Bibliography

Altman, Lawrence K. 'Complicated Surgery through Tiny Incisions.' *New York Times*, 14 August 1990.

Altman, Lawrence K. 'Surgical Injuries Lead to New Rule: New York Assails the Training for a Popular Technique.' *New York Times*, 14 June 1992.

Anderson, Julie, Francis Neary, and John Pickstone. *Surgeons, Manufacturers and Patients: A Transatlantic History of Total Hip Replacement*. Basingstoke: Palgrave Macmillan, 2007.

Anonymous. 'Atlanta Physician Pioneers Laser Gallbladder Surgery.' *Atlanta Daily World*, 25 January 1990.

Anonymous. 'CGH Uses Laser Surgery for Gall Bladder.' *Times Tribune*, 15 July 1990.

Anonymous. 'CONGRATULATIONS to Michael Gleeson...' *Scranton Times*, 24 October 1990.

Anonymous. 'Laser Gallbladder Surgery Cuts Time in Hospital, Recuperation.' *Clarksdale Press Register*, 18 December 1989.

Anonymous. 'Lasers Make Surgery Easier at Carbondale General.' *Scranton Times*, 20 July 1990.

Anonymous. 'Laser Surgery First in State.' *Greenwood Commonwealth*, 18 December 1989.

Anonymous. 'Methodist Pioneers Laser Gallbladder Removal.' *Hattiesburg American*, 25 January 1990.

Anonymous. 'Philippe Mouret, l'inventeur lyonnais de la coelioscopie, est mort.' *LyonMag*, 24 June 2008. www.lyonmag.com/article/8141/philippe-mouret-l-8217-inventeur-lyonnais-de-la-coelioscopie-est-mort. Accessed 2 July 2017.

Anonymous. 'Woman First in State to Undergo Innovative Surgery.' *Daily Press*, 29 January 1990.

Berci, George. 'Laparoscopic Cholecystectomy Viewed from the USA.' *Australian and New Zealand Journal of Surgery* 61 (1991): 249–50.

Billingsley, Anna Barron. 'New-Surgery Pioneer Is Model Patient.' *Richmond Times-Dispatch*, 27 January 1990.

Bor, Jonathan. 'New Technique Eases Gallbladder Surgery: Surgeons at UM Use Laparoscope.' *Baltimore Sun*, 27 October 1989.

Braun, Ramona. 'Laparoscopy as a Neo-Eugenic Practice in Gynaecology, 1940s–60s.' PhD dissertation, University of Cambridge, 2015.

Carbondale General Hospital. 'Another Medical First at Carbondale General Hospital.' *Scranton Times*, 29 November 1990.

Carbondale General Hospital. 'Which of These Women Had Gallbladder Surgery Yesterday?' *Scranton Times*, 7 June 1990.

Carbondale General Hospital. 'Which of These Women Had Gallbladder Surgery Yesterday?' *Press & Sun-Bulletin*, 7 June 1990.

Carbondale General Hospital. 'Which of These Women Had Gallbladder Surgery Yesterday?' *Tribune*, 8 June 1990.

Carbondale General Hospital. 'Which of These Women Had Gallbladder Surgery Yesterday?' *Ithaca Journal*, 11 June 1990.

Carbondale General Hospital. 'Which of These Women Had Gallbladder Surgery Yesterday?' *Citizens' Voice*, 16 June 1990.

Carbondale General Hospital. 'Which of These Women Had Gallbladder Surgery Yesterday?' *Times Leader*, 17 June 1990.

Carbondale General Hospital. 'Which of These Women Had Gallbladder Surgery Yesterday?' *Pottsville Republican*, 20 June 1990.

Carbondale General Hospital. 'Which of These Women Had Gallbladder Surgery Yesterday?' *Press-Enterprise*, 20 June 1990.

Carbondale General Hospital. 'Which of These Women Had Gallbladder Surgery Yesterday?' *Morning Call*, 25 June 1990.

Carbondale General Hospital. 'Which of These Women Had Gallbladder Surgery Yesterday?' *Daily Record*, 31 July 1990.

Carbondale General Hospital. 'Which of These Women Had Gallbladder Surgery Yesterday?' *Star-Gazette*, 5 August 1990.

Carson, Kelly. 'Doctor Zaps Gallbladder for 1st Operation in State: New Laser Surgery Saves Recovery Time.' *Hattiesburg American*, 17 December 1989.

Carson, Kelly. 'Laser Surgery May Be Taught Here.' *Hattiesburg American*, 8 January 1990.

Cuschieri, Alfred 'The Laparoscopic Revolution – Walk Carefully before We Run.' *Journal of the Royal College of Surgeons of Edinburgh* 34 (1989): 295.

Cuschieri, Alfred, George Berci, and Charles K. McSherry. 'Laparoscopic Cholecystectomy.' *American Journal of Surgery* 159 (1990): 273.

Desmarais-Tremblay, Maxime. 'W.H. Hutt and the Conceptualisation of Consumers' Sovereignty.' *Oxford Economic Papers* 72 (2020): 1050–71.

Emery, Margaret. 'Simplifying Surgery.' *Tribune*, 13 August 1990.

Epstein, Steven. *Impure Science: AIDS, Activism, and the Politics of Knowledge.* Berkeley, CA: University of California Press, 1996.

Escarce, José J., Judy A. Shea, and J. Sanford Schwartz. 'How Practicing Surgeons Trained for Laparoscopic Cholecystectomy.' *Medical Care* 35 (1997): 291–6.

Flannery, Mary. 'The Changing "Scope" of Surgery: New Technique Minimizes Scarring, Allows Faster Recovery.' *Philadelphia Daily News*, 6 June 1990.

Foretich, Chantel. 'Hub Doctor Led State in New Surgery.' *Hattiesburg American*, 24 July 1990.

Fried, Gerald. Interview by Thomas Schlich and Cynthia L. Tang, 13 June 2014.

Gollan, John L., Sarah Kalser, Henry Pitt, and Steven Strasberg. 'Foreward: Gallstones and Laparoscopic Cholecystectomy.' *American Journal of Surgery* 165 (1993): 387–8.

Graves, Herschel A., Jeanne F. Ballinger, and William J. Anderson. 'Appraisal of Laparoscopic Cholecystectomy.' *Annals of Surgery* 213 (1991): 655–62.

Green, Frederick L. 'New York State Health Department Ruling – A "Wake-Up Call" for All.' *Surgical Endoscopy* 6 (1992): 271.

Groves, Bob. 'New Light on Gallbladder Surgery: Doctors Embrace Laser Method.' *The Record*, 27 July 1990.

Hansen, Bert. *Picturing Medical Progress from Pasteur to Polio: A History of Mass Media Images and Popular Attitudes in America.* New Brunswick, NJ: Rutgers University Press, 2009.

Holmes, Judy. 'A Cut Above: Technique Cuts Recovery Time and Risk for Gallbladder Patients.' *Asbury Park Press*, 1 May 1990.

Hooven, Theodor. 'Gallbladder Article Timely.' *News Journal*, 2 July 1990.

Kalser, Sarah C. et al. 'National Institutes of Health Consensus Development Conference Statement on Gallstones and Laparoscopic Cholecystectomy.' *American Journal of Surgery* 165 (1993): 390–6.

Kern, Kenneth A. 'Medicolegal Analysis of Bile Duct Injury during Open Cholecystectomy and Abdominal Surgery.' *American Journal of Surgery* 168 (1994): 217–22.

Kornblau, Barbara. Interview by Cynthia L. Tang, 11 November 2021.

Labalme, Jenny. 'RMC Lasers Cut Surgery Time.' *Anniston Star*, 9 March 1990.

Lederer, Susan. *Flesh and Blood: Organ Transplantation and Blood Transfusion in Twentieth-Century America.* New York: Oxford University Press, 2008.

Lerner, Barron H. *The Breast Cancer Wars: Hope, Fear, and the Pursuit of a Cure in Twentieth-Century America.* Oxford: Oxford University Press, 2001.

Linker, Beth. 'Prosthetic Imaginaries: Spinal Surgery and Innovation from the Patient's Perspective', in *Technological Change in Medicine: Historical Perspectives on Innovation*, edited by Thomas Schlich and Christopher Crenner, pp. 100–28. Rochester, NY: University of Rochester Press, 2017.

Linsalata, Mark. 'Laser Shines in Surgery on Gallbladder.' *San Antonio Express-News*, 31 January 1990.

Litynski, Grzegorz S. *Highlights in the History of Laparoscopy: The Development of Laparoscopic Techniques – A Cumulative Effort of Internists, Gynecologists, and Surgeons.* Frankfurt/Main: Barbara Bernert Verlag, 1996.

McKernan, Barry. Interview by Cynthia L. Tang, 4 June 2018.

Meyer, Harris. 'Danger on the Cutting Edge?' *Los Angeles Times*, 29 July 1992.

Morgenstern, Leon. 'An Unsung Hero of the Laparoscopic Revolution: Eddie Joe Reddick, MD.' *Surgical Innovation* 15 (2008): 245–8.

Mouret, Philippe. 'Special Lecture: How I Developed Laparoscopic Cholecystectomy.' *Annals of the Academy of Medicine* 25 (1996): 744–7.

Mulhern, Barbara. 'New Gallbladder Surgery Technique Sharply Cuts Recuperation Period.' *Capital Times*, 12 June 1990.

Nathoo, Ayesha. *Hearts Exposed: Transplants and the Media in 1960s Britain*. Basingstoke: Palgrave Macmillan, 2009.

Naunton, Ena. ' "Chopsticks" Turn Rough Surgery into Easy Takeout.' *Miami Herald*, 16 March 1990.

Naunton, Ena. 'Doctors Call It "Nintendo Surgery": Gallbladder Operation Is One of the Hottest Surgical Advances in Years.' *Greenville News*, 22 March 1990.

Naunton, Ena. 'New Bellybutton Surgery Excises Sick Gallbladders.' *Hartford Courant*, 29 March 1990.

Naunton, Ena. 'New Type of Gallbladder Operation.' *Fresno Bee*, 27 March 1990.

Naunton, Ena. 'Now There's a Smaller Cut for Gallbladder Surgery.' *Morning Call*, 3 June 1990.

Nevada, Charlene. 'Gallbladder Surgery Technique Being Done in Ohio.' *Akron Beacon-Journal*, 13 February 1990.

New York State Department of Health. 'Laparoscopic Surgery.' New York State Department of Health Memorandum – Series 92–20, 12 June 1992.

Olsen, Douglas. Interview by Cynthia L. Tang, 29 May 2018.

Page, David W. *The Laparoscopic Surgery Revolution: Finding a Capable Surgeon in a Rapidly Advancing Field*. Santa Barbara, CA: Praeger, 2017.

Palko, Jerry. 'Surgical Technique Introduced in Region.' *Tribune*, 11 May 1990.

Perl, Rebecca. 'New Procedure to Remove Gallbladder Reduces Cost, Eliminates Hospital Stay.' *Atlanta Constitution*, 2 January 1990.

Prentice, Rachel. *Bodies in Formation: An Ethnography of Anatomy and Surgery Education*. Durham, NC: Duke University Press, 2012.

Ritter, Malcolm. 'Gallbladder Removal Avoids Incision, Speeds Recovery.' *Galveston Daily News*, 13 August 1990.

Roan, Shari. ' "Keyhole" Incisions Are Making Surgery Less Painful, Cheaper.' *Los Angeles Times*, 3 July 1990.

Sekellick, Elizabeth. 'New Hope for Gallbladder Patients.' *Star-Gazette*, 21 August 1990.

Sirgy, M. Joseph, Dong-Jin Lee, and Grace B. Yu. 'Consumer Sovereignty in Healthcare: Fact or Fiction?' *Journal of Business Ethics* 101 (2011): 459–74.

Skipper, Deborah. 'The Sky's the Limit for Laser Surgery Applications: Doctors at the Surgery Clinic of Hattiesburg Were the First to Perform Laser Laparoscopic Cholecystectomy.' *Clarion-Ledger*, 18 March 1990.

Tancill, Karen B. 'Surgeons Are Taking the Navel Route to Remove Gall Bladders. But It's No Gut Issue.' *Journal Times*, 14 March 1990.

Tang, Cynthia L. 'Physician Advertising and the "Patient-Driven Revolution" in Surgery, c. 1970–1990.' *Bulletin of the History of Medicine* (forthcoming).

Tang, Cynthia L. 'Technological Change in "Ordinary Medicine": The Emergence of Minimally Invasive Gallbladder Surgery, c.1970–1992.' PhD dissertation, McGill University, 2021.

Teveen, Joyce. 'Medicine Gets Less Incisive: Doctors Using Lasers to Remove Gallbladders.' *Argus Leader*, 22 May 1990.

Tomes, Nancy. 'Patients or Health-Care Consumers? Why the History of Contested Terms Matter', in *History and Health Policy in the United States*, edited by Rosemary Stevens, Charles E. Rosenberg, and Lawton Burns, pp. 83–110. New Brunswick, NJ: Rutgers University Press, 2006.

Twedt, Steve. 'Gallbladder Technique Used Here First Time.' *Pittsburgh Press*, 20 October 1989.

Vignal, Jacques. Interview by Cynthia L. Tang, 2 July 2018.

Walton, Richard D. 'New Method Eases Pain of Gallbladder Removal.' *Indianapolis Star*, 19 November 1989.

Whitfield, Nicholas. 'A Revolution through the Keyhole: Technology, Innovation, and the Rise of Minimally Invasive Surgery', in *The Palgrave Handbook of the History of Surgery*, edited by Thomas Schlich, pp. 525–48. London: Palgrave Macmillan, 2018.

Winslow, Ron. 'Cutting Edge: A Tiny TV Camera Is Fast Transforming Gallbladder.' *Wall Street Journal*, 10 December 1997.

Zetka, James R. *Surgeons and the Scope*. Ithaca, NY: Cornell University Press, 2003.

Part II

Informed patients and patient information

4

Tampons, technology, and toxic shock syndrome: From consumer to patient to informant

Sharra Vostral

In 1980 the media alerted women to a new and frightening illness associated with menstrual periods and tampon use: toxic shock syndrome (TSS). The illness raised grave concerns because it struck healthy individuals and because 70 per cent of women used tampons. With so many people at risk, the fear was that many might get sick or even die if the problem was not contained swiftly and effectively. The numbers did not prove to be as catastrophic as imagined, with the United States Centers for Disease Control and Prevention (CDC) confirming 941 cases and seventy-three deaths between 1970 and 1980.[1] Still, many public health officials and epidemiologists argued that the official numbers were low due to the passive data collection system used by the CDC, as well as the difficulty of meeting the strict criteria of a clinical case definition, in which six of six categories must be met before the diagnosis was 'confirmed' (Figure 4.1). This limiting definition excluded many women who presented insufficient symptoms to be counted among the confirmed cases. Early symptoms of TSS deceptively resembled the flu, but quickly deteriorated into septic shock and death if not quickly treated with antibiotics.

The elusive origins and complex presentation of TSS resisted a simple explanation.[2] Part of the reason is the complexity of the disease process – unknown at the time – which is determined by several variables including the tampon, the individual woman's microbiome, and her immune system. However, the emergence of new super-absorbent tampons in the late 1970s and early 1980s was a key factor. Among them was the Rely brand manufactured by Procter & Gamble (P&G) that departed from a traditional cotton composition which dated back to the 1930s to one that was 100

Fever (102 F/39 C)

Rash

Desquamation (flaking, peeling skin)

Hypotension (drop in blood pressure, dizziness)

Multisystem involvement, with three or more of the following:

- Gastroentestinal distress (vomiting, diarrhoea)
- Muscular pain (creatine phosphokinase levels twice that of normal)
- Mucus membranes (enlarged blood vessels within eye, throat, or vagina)
- Renal dysfunction
- Liver dysfunction
- Blood abnormalities
- Central nervous system issues

Negative cultures

- Rocky Mountain spotted fever
- Blood and cerebrospinal fluid

Figure 4.1 Toxic shock syndrome: clinical case definition; clinical criteria as outlined by the Centers for Disease Control

per cent synthetic.[3] They were made of a polyester shell, polyester foam cubes, and an absorbent gel material called carboxymethylcellulose. When the tampon became moistened, it was designed to open into the shape of a cup, which additionally served as a vessel to catch and hold menstrual fluid. For women with heavy periods, it worked wonders. But women's bodies are variable, and for others it served as a catalyst of illness. These women harboured the bacterium *Staphylococcus aureus* in their vaginal microbiome and never developed the necessary antibodies to counteract the toxins the bacterium produced. The super-absorbent tampon additionally introduced oxygen into a usually anaerobic vaginal canal, altering the environment for bacterial growth. Facilitated by less acidic pH during menstruation, the conditions were favourable for unchecked growth and toxin production that precipitated septic shock.[4]

Epidemiological studies conducted by the CDC subsequently determined that all super-absorbent tampons carried a risk for TSS, but the Rely tampon showed the highest rate of all. Procter & Gamble publicly received accolades because the company voluntarily withdrew its product and alerted women to the dangers of Rely. This, however, was a façade that covered a less ethical practice. Behind the scenes, the corporation fought for access to the women's anonymized data from the CDC studies. This chapter examines the 1985 lawsuit in which Procter & Gamble sued the CDC for the right to identify the women and access their medical records. This was complicated by their overlapping identities of: 1) consumer, through the purchase of manufactured tampons; 2) patient, through the illness of tampon-related TSS due to product use; and 3) informant, in epidemiological studies of an emergent illness. Each category carried more or less weight for different stakeholders, but the sickened women embodied them all. The patient/consumer paradigm assumes a level of personal agency; however, these particular identities (in this case) were not voluntarily made. Moreover, the prevailing belief system of menstrual stigma draped a cloak of shame over the entire proceedings. Menstrual stigma affected consumers in the mandatory purchase of products to absorb and hide a disparaged bodily fluid. Menstrual stigma affected the communication of risk and how quickly patients should access timely treatment. Menstrual stigma affected informants and their right to remain shielded from menstrual embarrassment. And, finally, menstrual stigma played a determining role regarding the long-term legal protections for women related to anonymity, confidentiality, and privacy as consumers, patients, and informants.

Background

Menstruation and its management, through the technologies of tampons and menstrual pads, serves as a useful case study for understanding the relationship between patients and consumers. This is unusual because menstruation is a normal bodily process not requiring medical intervention, so there are no patients involved. Through the commodification of menstrual products that women purchased, they became consumers in relation to menstruation. It was only because of the purchase and use of certain tampons, however, that some women became patients through product injury. This injury was imprinted with the stigma of menstrual fluid and the menstrual process itself. Despite menstrual products being robustly advertised and intensively marketed in newspapers, maga-zines, and on television, menstruation and menstrual blood was, and continues to be, conspicuously absent. Its lack of visual repre-sentation gestures to the deep cultural beliefs of stigma and shame surrounding it.[5]

In Western thought, menses has been historically identified as a biological marker of women and womanhood, and therefore the opposite of *man*, who is considered rational and in possession of personal political autonomy.[6] Menstruation has been used as a rationale for limiting the political status for women based on Enlightenment and natural rights politics.[7] Today, menstruation is not weaponized in quite the same way, but it still operates through stigma to establish limitations on women and their bodies.[8] In my book *Under Wraps: A History of Menstrual Hygiene Technology* (2008), I conceptualized twentieth-century menstrual products as 'technologies of passing' to describe the process by which they help women to overcome social and political restrictions.[9] Using men-strual technologies to pass as non-menstrual helps to circumvent embarrassment from leaks and skirt historical prohibitions during menstruation, such as swimming, or to mitigate menstrual absences that damage workplace efficiency.[10] Under the guise of 'health' and 'hygiene' and even 'wellness', tampons as medical technologies have been promoted to improve the troublesome menstruous body. This premise that menstruation is problematic is so culturally pro-found that it also informed the outcome of the Procter & Gamble litigation. The ruling protected women's data, including informa-tion about menstrual cycles, sexual partners, sexually transmitted infections, and douching practices, for the reason that it would be embarrassing to women to expose this information.[11] The verdict in favour of the CDC hinged upon the expectation for informants (the women) to retain anonymity and avoid embarrassment, and not the more meaningful right to medical privacy or equal pro-tection as United States citizens.[12] Thus, both the consumer tech-nologies and the newly established court precedent of anonymity left stigma unquestioned and reinforced passing and socially con-structed agreements about propriety. What began with the tech-nological use of tampons transformed women from consumers, to patients, to informants requiring the court's protection from public embarrassment.

Consumer patient

The transformation of the sick person seeking medical care – identified as a patient – to an ill person making a choice about and paying for a particular treatment – as a consumer – has shaped medical care in the United States. Both Alex Mold and Nancy Tomes outline the contours in their respective works, detailing the fusion of the patient consumer in the twentieth century.[13] This volume also emphasizes this problematic transformation of the patient consumer (as discussed in the introduction), while including technology as an important co-factor in the purchase and reception of treatment. I choose to examine the reverse: the case when a consumer becomes an unanticipated patient. If there is a degree of choice exercised in the role of consumer, then the experience of the patient can be characterized as involuntary. However, there are many cases of individual consumers becoming patients due to technological injury. Sarah Lochlann Jain, an anthropologist whose work falls at the intersections of science, technology, gender, and medicine, details examples from product design injury, and describes the calculations that made corporations ignore intrinsic flaws of their product for the sake of cost-effectiveness. The Ford Pinto and its exploding engine, or the case of carpel tunnel syndrome from poor keyboard design, demonstrate corporate choices of overlooking problems in which human injury is understood as an 'acceptable' part of the cost of doing business.[14] Pollutants and environmental contamination are additional examples of this kind of rationale. The unsuspecting users of contaminated drinking water, air, or food suffer from toxic exposures and thus enter the category of accidental patient.[15] Similarly, Allan Brandt, a historian of medicine, has outlined the toll of tobacco-related diseases, which is an additional cost paid by a consumer becoming a patient suffering from lung cancer, or an unwilling breather suffering the effects of second-hand smoke.[16] Others such as Naomi Oreskes and Erik Conway further illuminate the deliberate withholding of information, and the manufacture of misinformation, by the tobacco industry to perpetuate doubt and maintain cigarette sales, keeping consumers purposely uninformed despite the deleterious health outcomes of smoking.[17] Reproductive technologies such as the Dalkon Shield intrauterine device or Norplant birth control implants also disproportionately injured many women of colour and low-income women who bore the deleterious consequences.[18] In these cases, healthy women became patients and even plaintiffs when corporate profit outweighed injury and irrevocable damage to their bodies.

These examples of consumers becoming patients include an element of the unknown. They unknowingly drove a car with fatal engineering flaws, or drank water contaminated with lead due to ageing pipes and changes in water treatment chemistry. When key knowledge is deliberately withheld from individuals, cultivating purposeful 'not knowing' or agnotology, they become patients involuntarily due to a lack of information or consent. Of course, being a patient is generally not a desirable category or even voluntary, but there is a particular kind of deception related to these manufactured items and technologies, as compared to contracting a communicable disease. Tampon users who experienced toxic shock syndrome also became such accidental patients. As consumers, they purchased over-the-counter super-absorbent tampons that caused harm indirectly by catalysing the bacterium *Staphylococcus aureus* to flourish when it otherwise would not, developing into TSS. The tampon was not a risk in and of itself or the direct cause of harm, but was a co-factor with the bacterium in precipitating illness and even death.[19]

The case of TSS offers one additional role in this relationship between consumer and patient, and that is informant. This chapter extends previous research on the social history of TSS by focusing beyond patients and even litigants, to the precarious role of the informant related to medical device injury. In 1980, when epidemiologists were working to identify the sources and causes of TSS, they enrolled women who had recovered from TSS as informants to their studies. This new identity also transformed their illness into data from which researchers drew conclusions and made recommendations about TSS and tampons. These consumers with presumed agency thus became accidental patients, and then obligatory informants. It was a role that they would not have willingly chosen, which in this case raises the question of the degree of free will that is involved in the patient consumer definition. The illness of tampon-related TSS expands the meaning of the patient consumer and is reflective of a contradiction concerning agency that exists more broadly within this construct.

Menstruation and medicalization

Menstruation has many cultural interpretations, but for purposes here it is useful to look back to the late nineteenth century, and the rise of modern medicine. Many of the newly introduced medical practices and theories had a beneficial impact on human health, for example, the bacteriological knowledge underlying practices like handwashing or the surgical technologies that enabled new operations for cancer.[20] At the same time, human conditions or behaviours without clear biological underpinnings also became subject to medical interpretation and were thus 'medicalized'.[21] Women's normal biological processes were particularly subject to medicalization, including menstruation, pregnancy, and menopause, as feminist historians have found.[22] Menstruation as a life-course event for most women was a strong contender for medicalization in the United States.[23] Even regular menstrual cycles were understood as problematic, disease-producing, and a threat to 'civilization' because they supposedly predisposed White women to be weak and feeble according to eugenicist logic. As male physicians in the late nineteenth century claimed expertise for treating women's so-called ailments, menstruation became understood in terms of a disease, disability, infirmity, and incapacity.[24] It is worth noting that this was a racialized and classed understanding; unhealthy menstruation affected privileged White women, while working-class labourers and enslaved women were seen as more 'natural' and less evolved, and therefore unable to experience menstrual ailments or even pain.[25] This pathologizing required further medical surveillance, and interventions were required to reclaim White women's health.

The sale of menstrual products in the 1920s tapped into these tropes, and created a new group of consumers based on the purported premise that menstruation was shameful, unhygienic, and indelicate, with the technologies providing the solution to transform women's bodies to clean and acceptable.[26] Consuming menstrual products also became a means to demonstrate modernity rather than remaining 'old-fashioned'.[27] It was a way for women to assert independence and claim bodily freedom through a one-time-use product. Women could relinquish the time-consuming duty of hand laundering and air drying reusable menstrual cloths, and instead purchase a disposable item to then burn, bury, or throw out in the garbage. The expansion of the market economy through the purchase of readily disposable and replaceable technologies, and the discarding of them, marked a growing affluence of the population. It also signalled new norms around bodily management.[28] These disposable purchases appealed to what Danya Glabau, a medical anthropologist and scholar of Science and Technology Studies, identifies as the 'hygienic sublime', the idea that the pursuit of cleanliness through consumer goods, coupled with prescribed behaviour, will create a transformative purity of the body.[29]

Thus, disposable menstrual products are more than just commodities. They are technoscientific tools to manage menstruation, becoming a 'technological fix' for a stigmatized bodily process to hide and contain a leaky body.[30] Technoscience is a hallmark of biomedicine, yet we generally do not associate technoscience with gendered technologies or stigmatized bodily processes. When innovations in medical science and engineering during the 1970s and 1980s changed the composition of some tampons and they were marketed as 'super-absorbent', many women happily purchased them, motivated to try the 'new and improved' consumer objects. For many menstruators with heavy periods, they were a technological solution to intercept and transform their leaky bodies. The tampons' design that unfurled into a cup, the super-absorbent synthetic components, and the potential for longer wear contributed to protection from exposure to leaks as well as the ruse of the pass.

The intertwining of menstruation, medicalization, consumerism, and technology in 1980 was the height of scientific progress, but it also had the capacity to create unwitting patients through technological injury. Tampons, a tool of technoscientifically managing menses, became vectors of the deadly illness TSS. This meant that some women quickly transitioned from consumer to patient, with trips to doctors' offices and emergency rooms, suffering from fevers and sepsis.

Ascertaining the origin of this illness required both defining the parameters of a new syndrome as well as coordinating epidemiological surveys to determine its scope. State departments of health in Wisconsin and Minnesota took the early lead in tracing the first unconfirmed cases of this new illness and worked in close coordination with the CDC in 1980. The CDC subsequently ran two studies, blandly named CDC-1 (27 June 1980) and CDC-2 (19 September 1980) and reported in the *MMWR* (*Morbidity and Mortality Weekly Report*). The first confirmed that tampons were associated with TSS, and the second ascertained which tampons were most likely to be involved in the illness.[31] It was in this context that these accidental patients also doubled as informants. Through detailed interviews, the CDC gathered information from women who were ill and recovered in order to track outbreaks and identify co-factors of the illness. While the studies determined that all super-absorbent tampons carried risk, the Rely tampon manufactured by Procter & Gamble demonstrated the highest rate of all. The company bore the brunt of the media attention and faced the threat of a Food & Drug Administration (FDA) recall. Procter & Gamble voluntarily suspended sales of Rely tampons and withdrew them from stores in September 1980.[32]

These patient informants, however, unwittingly took on a particularly vulnerable and contested role. While Procter & Gamble appeared to take responsibility for its product and alerted women through newspapers, radio, and television public service announcements about the dangers of Rely, these actions deflected attention from more inauspicious intentions to apprehend the data collected by the CDC. Access to data became the ultimate goal for Procter & Gamble executives and lawyers. Procter & Gamble pressured, challenged, and sued the CDC for the right to identify informants and access their medical records. The company had strong commercial interests in seeking to retrieve the data that the CDC collected, presumably to discover epidemiological errors and ultimately clear its name. It proceeded to subpoena records, which the CDC contested, and resulted in *Farnsworth v. Procter & Gamble v. Center for Disease Control* (1985), which ultimately provided protection to the CDC data. However, it was the women and their intersecting identities of consumer, patient, and informant that swayed the appellate judges. It was not product liability as consumers through the US Medical Device Amendments (1976), or privacy as patients, but the right to be free of embarrassment that moved the court. The politics of menstrual stigma carried more weight than any of the prior identities in offering legal protections.

Tampons, big business, and consumerism

The 1980s marked a shift in attitude in many things: disco to Madonna, straight hair to big hair, and regulation to deregulation of neoliberal corporations. In fact, the ethos of Milton Friedman, the free-market economist, informed an attitude among some businesses that the ethical commitment that they had was to stockholders, and not necessarily to their own community of workers, consumers, and nearby neighbours. This is known as 'stockholder theory'. With stockholders their primary loyalty, the well-being of these other groups was of less, and even no, concern. This shift away from providing for employees and local communities resulted in a growing wage gap between hourly labour and management, with CEO salaries supplemented with handsome stock options as rewards for increasing corporate value. It also created an environment in which lawsuits, filed ultimately on behalf of the shareholders, became a necessary means and moral imperative to protect corporate earnings. Businesses had a duty to seek out and engage in litigation when protecting the interests of their shareholders.

Procter & Gamble was no exception to this business thinking. If raising corporate value for shareholders was the ultimate goal, then the investment of years of research and development into a tampon that was later withdrawn under the threat of a federal recall jeopardized corporate profitability. It harmed stockholders. For Procter & Gamble, it was vital and necessary to understand why the profitability of its product Rely was not only undermined by the FDA but also damaged by hundreds of product liability lawsuits leading to further losses. The key to this was the epidemiology conducted by the CDC that identified Rely as the most dangerous and deadly of all tampons, the verdict of which killed a product line upon whose viability and growth Procter & Gamble had invested capital. From this perspective, and because Procter & Gamble believed it had conducted more than sufficient testing, leaders in the company presumed the FDA conclusion was wrong, and therefore the data must be wrong as well. The best way to exonerate itself was not only to obtain the exact data but also to follow up with the same women who provided it to the CDC.

During the summer and autumn of 1980, Procter & Gamble embarked on an all-out evidence collection campaign, leaving no stone unturned. The goal was to locate and talk to the women who had purportedly fallen ill after using Rely tampons, and to their physicians who cared for them, in order to obtain access to their medical history and records. Procter & Gamble deployed three main tactics to access this information. First, they culled through all the calls intercepted by phone receptionists to the customer service phone number, looking for women to call back who had complained about feeling ill after using Rely. Second, if women happened to write to the company about their illnesses, this provided an opportunity for product managers to respond back to them and learn about their symptoms and tampon use.[33] For example, according to Roscoe Owen Carter, the PhD chemist in charge of paper products development and therefore also of Rely, who testified in a product liability trial, 'the only way that you could make a decision as to whether this might have been toxic shock syndrome was to get to the physician, talk with him, and then actually see the medical records, [and] go through these medical records'.[34]

In fact, Carter had already deployed this tactic with a woman in Indiana and her daughter. The mother purchased Rely tampons from the local Kmart that her daughter subsequently used. She reported to Procter & Gamble on 25 July 1980 that her doctor believed her teenage daughter was hospitalized with a staphylococcus infection caused by Rely.[35] Carter directly called the Kmart store manager, who presumably knew both due to the small size of the town and provided him with the identity of both the doctor and the patient. Carter subsequently contacted the physician, who shared that the teen experienced symptoms ranging from high fever, to muscle pain, to diarrhoea. While a store manager was not a medical professional bound to protect patient confidentiality, he nonetheless acted less ethically than one would hope. Under the guise of collecting medical evidence, strategists at Procter & Gamble flagrantly violated patient confidentiality.[36]

However, these in-house pursuits only went so far. Procter & Gamble needed aggregate data and access to the patients at ground zero: the informants participating in CDC-1 and CDC-2 conducted during the spring and summer of 1980.[37] These studies delivered evidence to the FDA about injury brought about by a medical device, so that the newly enacted Medical Device Amendments of 1976 could be used to issue a product recall. It was Procter &

Gamble's belief that women never used their tampons, were mis-diagnosed, or the CDC epidemiology was simply erroneous and conducted poorly.[38] Whatever the case, Procter & Gamble's third tactic was to have its own scientists obtain and examine the data and interview the informants from those original CDC studies. In this way, they sought to reclaim the tampon, the brand, the company, and profit for shareholders.

Medical health law

In the meantime, the CDC had experienced what could be called a 'crash course' in managing and containing a unique health crisis related to tampon technology. As I argue in my book *Toxic Shock: A Social History*, this was one of the first times that a technology was recognized as a *co-factor* in the production of an illness (unlike direct injury caused by a faulty automobile airbag or the Dalkon Shield). It was a high-profile case, and the Epidemic Intelligence Service (EIS), composed of skilled physicians and scientists at the CDC, provided the expert voices and media contacts for disseminating information to the public about TSS and the risks of tampon use. A hallmark of their approach was cooperation with stakeholders, including all the tampon manufacturers in the country, to share data and exchange information and knowledge about a rapidly worsening condition. Thus, in the aftermath of the product withdrawal, when Procter & Gamble requested the data from the studies, the CDC provided 34,000 pages and documents related to their research, with only names and addresses redacted.[39] This direct request and document delivery to the corporation bypassed usual Freedom of Information Act (FOIA) paperwork and its inevitable delays to foster expedient and cooperative corporate-government problem solving about TSS. Procter & Gamble, however, felt that the redactions were not in the spirit of the request, though disclosure of medical records through the FOIA is invasion of personal privacy and exempted in the act. The circumvention around the FOIA likely increased the amount and types of information shared with Procter & Gamble, but for lawyers there, the incomplete documents hampered their ability to interview the women and conduct an accuracy check of the CDC research.

Gene Matthews, the chief legal advisor to the CDC from 1979 to 2004, recalled: 'We were left on our own to deal with this trench warfare that went on for about five years.'[40] He and the legal team composed of Karen Kaunitz and Verla Neslund from the CDC, Nancy Buc from the FDA, and Nina Hunt, the Assistant US Attorney General of the Northern District of Georgia, rallied to defend the CDC from a subpoena filed by Procter & Gamble in 1981 for the epidemiologic records to include names and addresses. As Matthews put it, 'That wasn't going to happen. We weren't going to give on that.' They filed a protective order in 1982 to challenge the subpoena in the US District court, but this was new territory. This was a first, and Matthews continued: 'no case ... had ever been litigated like this where a company was wanting the names of women ... of subjects in an emergency outbreak public health investigation'.[41] At issue was the lack of a statutory basis to deny Procter & Gamble's request. There was a common-sense argument that anonymity was necessary for good public health research because private information was relayed to epidemiologists. In this case, questions addressed sexual histories, douching practices, menstrual cycles, and STIs, covering a broad swath of co-factors as they searched for and eliminated variables related to the illness. The informants, some of whom were teenagers, trusted that this information would be kept confidential due to social repercussions.

Matthews took a proactive approach and sent letters to all the informants, explaining that Procter & Gamble was interested in talking to them about their TSS experience, and asking whether they would want to be contacted by Procter & Gamble for such a purpose. He received the following responses: thirty-two agreed to the disclosure, 119 declined, and twenty-six were inaccurate addresses or simply not returned.[42] This proved useful in court, because it offered evidence of women's unwillingness to have their identities disclosed to Procter & Gamble.

As Matthews put it, 'Before you go to the mat on litigation, you've got to decide, can you afford to lose?' This question took on extra importance because the illness of HIV/AIDS had just emerged within the CDC lexicon of epidemics. TSS provided a dress rehearsal for the EIS, the spokespersons, and the legal counsel, about the demands that HIV/AIDS would levy, especially in terms of confidentiality of informants about this virus. This, though, also provided Matthews with some reassurance. He reasoned:

> We began to believe if we got a court opinion that ruled against us, we had a pretty good chance of going back to congress and making a compelling argument that this [a legal statute] was needed ... if you want us to be able to get valid information in order to protect you from disease outbreaks like AIDS that was [sic] going on.[43]

HIV/AIDS proved to be far more deadly, stigmatizing, and broad-reaching than tampon-related TSS, with confidentiality ramifications reverberating through schools, the workplace, and insurance coverage. The CDC legal team felt that the implications of protecting its TSS research would determine whether it would continue to conduct quality epidemiology, and be a trusted, effective, and crucial federal institution.

Although there were four separate oral arguments that the CDC legal team prepared, for the purposes here I will focus on the final case brought to the Eleventh Circuit US Court of Appeals in 1985, better known as *Farnsworth v. Procter & Gamble v. Center for Disease Control*.[44] The appeal for Procter & Gamble hung on the legal precedent of 'discovery', that is, the company had the right to discover the names of the informants in the CDC studies in order to verify or find errors in the epidemiology. According to the court, 'the law's basic presumption is that the public is entitled to every person's evidence' and that 'the Federal Rules of Civil Procedure strongly favour full discovery whenever possible'.[45] There was a tension between the tenet of freedom of information and the need to weigh business interests equally with that of the individual, and in this case 'to protect a party or person from annoyance, embarrassment, oppression, or undue burden or expense'. The court found that 'Under that standard, the district court's duty was to balance Procter & Gamble's interest in obtaining the names and addresses of the study participants against the Center's interest in keeping that information confidential'.[46]

It is useful to pause here and reflect upon subtle differences in language and terminology. 'Anonymity' allows one to be anonymous and not known. This differs from 'confidentiality', which is an agreement between at least two parties that something be kept secret, or not disclosed. Lastly, privacy is the legal right to not have one's personal matters disclosed or publicized. This is especially true of government intrusions in personal issues. Privacy is contextual, and Eden Osucha, who specializes in American literature, culture, and legal studies, argues that privacy is inherently defined through Whiteness, going further to say that privacy is also a form of property.[47] White women were the dominant demographic of the studies, and their property and privacy were interrelated, with private medical information – in this case data – a form of property. In essence, the women relinquished their rights to their personal data, their property, for the good of the state. But did the CDC have the sole right to this private property? The slippage and manipulation of language by both Procter & Gamble and the CDC regarding anonymity, confidentiality, and privacy shows the very unstable and problematic category of rights for women informants and who owned their data.

It was clear, however, that women were not interested in Procter & Gamble gaining free access to their data. Matthews' efforts of outreach and his mini survey about women's desire for confidentiality proved to be a significant influence on the court. It demonstrated the CDC's effort to work with the company in good faith, and even deliver the details of thirty women who agreed to talk to them. The court discerned that 'P&G [Procter & Gamble] also has the identity of at least fifty women involved in the study. They have apparently cooperated with Procter & Gamble and biases of the study participants may well be discovered from questioning these women'.[48] However, Procter & Gamble sought more access. Matthews described one moment in the oral argument when Larry Elleman, one of the Procter & Gamble lawyers, tried to explain to the panel of all-male judges why they wanted to contact the women 'just to ask them questions and see if this [the science] is right'. Matthews recalled that one of the judges 'roused up, and leaned forward and glared down and said, "Haven't the women already said they don't want to talk to you?"' He continued: 'I was feeling a little better that day ... we won, and we won big'.[49] Procter & Gamble pursued this matter no further with a higher court.

This win included a validation of the CDC's work and ability to operate within an ethic of guaranteeing anonymity. The court determined:

> First, the Center's purpose is the protection of the public's health. Central to this purpose is the ability to conduct probing scientific and

social research supported by a population willing to submit to in-depth questioning. Undisputed testimony in the record indicates that disclosure of the names and addresses of these research participants could seriously damage this voluntary reporting. Even without an express guarantee of confidentiality there is still an expectation, not unjustified, that when highly personal and potentially embarrassing information is given for the sake of medical research, it will remain private.[50]

Though we are now accustomed to Institutional Review Board (IRB) oversight, it was only in 1974 that the National Research Act established IRBs to review medical and behavioural research in the United States. The lawyers at the CDC were very much at the forefront of establishing protocol regarding the 'highly personal' and 'potentially embarrassing' information about informants, as problematic as that conceptualization was.

Implications

This 1985 ruling by the appellate court was an important victory for the CDC, noted by William Curran, a legal specialist in health law, who had a long-running notes section on medical intelligence in *The New England Journal of Medicine*. He interpreted the case more directly for practitioners and concluded in his commentary in 1986 that 'In one important aspect, the CDC has achieved an important victory for all researchers who ensure the confidential-ity of personal identity in voluntary epidemiologic studies that are conducted in the interest of the public's health.'[51] But he had a far more startling assessment, one that also becomes more trouble-some when interpreted through a feminist lens. Curran surmised that 'The court, in fact, asserted at the end of its opinion that its decision did "not depend upon any legal privilege"', meaning that 'the court was not relying on any other legal-ethical privilege of professional confidentiality to protect the names and addresses of the subjects of the study'. As Curran put it, 'These women were not patients to whom an obligation of confidentiality is tradition-ally owed by physicians.' They were merely 'revealing information about their personal lives and experiences to an investigator for a federal agency'.[52]

Thus, if women in the study were not patients, all the usual professional agreements about the Hippocratic Oath, the Nuremberg Code, and the AMA Code of Ethics (1847)[53] did not apply. The ruling protected the CDC and its ability to carry out the business of public health, but not women's rights to their bodily privacy. Here, the court maintained its long history of paternalistic protectionism while denying basic rights of citizenship to women. For instance, in *Griswold v. Connecticut* (1965), the right to birth control was guaranteed only to married couples and rationalized as a private decision between a husband and wife that the government could not infringe upon. Single women gained the right to contraception in 1972 with *Eisenstadt v. Baird* because it violated the equal protection clause. Privacy was a prevailing factor in *Roe v. Wade* as well, in which both the doctor's and patient's right to privacy, and the doctor's right to practise medicine without infringement, provided logic for the ruling which has more recently been overturned by *Dobbs v. Jackson Women's Health Organization* (2022).[54] Privacy did not provide the means to protect abortion rights.

Hinging rights on privacy is flawed. Heidi Schreck, a playwright, discusses how women have been left out of the US Constitution, and paraphrases Supreme Court Justice Ruth Bader Ginsburg's interpretation of the use of privacy as problematic to women's human rights. She says:

> Ruth Bader Ginsburg argued that if the right to reproductive freedom could have been based in equal protection under the law, with the idea that you can't possibly be equal in this country, as a woman, if you don't have bodily autonomy, if you don't have the right to decide what to do with your own body – the right to decide whether to have children or not – if childbearing is obligatory, then you can't possibly be equal in this country.[55]

The takeaway from Bader Ginsburg's interpretation is that the Fourteenth Amendment to guarantee equal protection under the law is not granted uniformly to women. Furthermore, while privacy is a central component of equality, it is also subject to vagaries related to who a person is; it is not an inalienable right.

Lurking behind decisions about women's privacy was the dark shadow of embarrassment. The premise of the right to confidentiality in *Farnsworth* was not privacy but the ostensible prevention of embarrassment to the informants. The presumption was that women might be ashamed of many things: their menstrual cycle, sexual activity outside the cultural bounds of marriage, sexual activity as teenagers not condoned by their parents, and sexually transmitted infections. It was not product liability as consumers or privacy as patients that moved the court, but the right to be free of embarrassment, as entangled with White middle-class gendered respectability, that swayed the appellate judges. The ruling agreed that embarrassing information should remain confidential and did not extend the full rights of legal privacy to women. Embarrassment operated on two fronts. The CDC now controlled the right to their private data, with legal conceptualizations of embarrassment institutionalized and protecting that information. And, sacrificing Rely tampons for the sake of all the other brands and sizes continued to protect women from the embarrassment of exposed menstrual leakage, even though TSS remains a risk of use.

Despite these structural shortcomings, at this moment in the 1980s, it was important for the CDC to accrue this win. Race, gender, and class continued to shape political, medical, and legal conceptions of whose privacy was worth protecting and whose should be disregarded in the name of the public good.[56] This held significant ramifications because it previously had no legal protections in place for people enrolled in HIV/AIDS studies in the early 1980s, contiguous with the TSS outbreak. TSS served as a precursor of sorts, and the *Farnsworth* win provided the right to anonymity more broadly to these epidemiological studies.

Conclusion

The illness of tampon-related TSS is a useful example to interrogate the construct of the patient consumer. This instance presents overlapping identities, and the involuntary nature of the consumer who becomes a patient. Moreover, the technological injury resulted in the unwanted identity of informant in epidemiological studies. This chain of events also unexpectedly precipitated a critical court ruling regarding the right to anonymity in studies related to federally conducted public health investigations. At the core was an assumption of confidentiality held by the women who contracted

TSS that the epidemiologists would uphold. In a system that was constantly challenged by commercial and outside interests and that did not necessarily honour individuals' rights as a core value, the CDC resorted to court proceedings to secure the women's right to anonymity. However, this right is still tenuous because it relies on White women's respectability, and their need to be protected from embarrassment and the shame related to pre-marital sex, STIs, and menstrual stigma. The premise of the court decision securing anonymity forewarns about the traps of privacy when justified through this guise. This has relevance to the contemporary moment when corporate entities broker individuals' digital health information data as a lucrative resource and commodity.[57] This focused examination of TSS demonstrates how the medical, legal, and social understandings of the categories of consumer, patient, and informant – by way of technology and menstrual stigma – have consequences for our collective health, and how data requires new conceptualizations of privacy to ensure consent and intent of use.

Notes

1 Jeffrey Davis, 'The Investigation of TSS in Wisconsin and Beyond, 1979–1980', in *Outbreak Investigation around the World: Case Studies in Infectious Disease Field Epidemiology*, edited by Mark Dworkin, pp. 79–201 (Boston, MA: Jones and Bartlett, 2009).
2 Sharra Vostral, *Toxic Shock: A Social History* (New York: New York University Press, 2018).
3 Sharra Vostral, *Under Wraps: A History of Menstrual Hygiene Technology* (Lantham, MD: Lexington Press, 2008).
4 Philip Tierno, *The Secret Life of Germs: Observations and Lessons from a Microbe Hunter* (New York: Pocket Books, 2004).
5 Camilla Mørk Røstvik, 'Crimson Waves: Narratives about Menstruation, Water, and Cleanliness.' *Visual Culture and Gender* 13 (2018): 54–63. http://vcg.emitto.net/index.php/vcg/article/view/114/118. Accessed 10 October 2023.
6 People who menstruate, women, and transgender folks are more frequently being referred to as menstruators. Not all cisgender women have periods either, so menstruator captures this bodily process. However, the historical category of *woman* has enrolled all people who menstruate, and this designation is currently being reimagined in terms of gender.

7 Denise Riley, *Am I that Name? Feminism and the Category of 'Women' in History* (London: Macmillan, 1988).

8 Margaret Johnson, 'The Period Is Political: Menstrual Justice, Abortion Rights, and Reproductive Justice', Keynote address for the bi-annual meeting of the Society for Menstrual Cycle Research, Washington DC, 2023. This previews her forthcoming work on surveilling menstruators.

9 Vostral, *Under Wraps*; on the menstrual hygienic imperative, see Joan Jacobs Brumberg, *The Body Project: An Intimate History of American Girls* (New York: Vintage Books, 1998); on periods and modernity, see Lara Freidenfelds, *The Modern Period: Menstruation in Twentieth-Century America* (Baltimore, MD: Johns Hopkins University Press, 2009).

10 Vostral, *Under Wraps*, Chapter 5.

11 William Curran, 'Protecting Confidentiality in Epidemiologic Investigations by the Centers for Disease Control.' *New England Journal of Medicine*, 314, no. 16 (1986): 1027–8.

12 *Farnsworth v. Procter & Gamble Company v. Center for Disease Control*, 758 F.2d 1545 (11th Cir. 1985). https://law.resource.org/pub/us/case/reporter/F2/758/758.F2d.1545.84–8330.html. Accessed 10 October 2023.

13 Alex Mold, *Making the Patient-Consumer: Patient Organisations and Health Consumerism in Britain* (Manchester: Manchester University Press, 2015); Nancy Tomes, *Remaking the American Patient: How Madison Avenue and Modern Medicine Turned Patients into Consumers* (Chapel Hill, NC: University of North Carolina Press, 2016).

14 Sarah Lochlann Jain, *Injury: The Politics of Product Design and Safety Law in the United States* (Princeton, NJ: Princeton University Press, 2006).

15 Phil Brown, *Toxic Exposures: Contested Illnesses and the Environmental Health Movement* (New York: Columbia University Press, 2007).

16 Allan Brandt, *The Cigarette Century: The Rise, Fall, and Deadly Persistence of the Product that Defined America* (New York: Perseus Books Group, 2007).

17 Naomi Oreskes and Erik M. Conway, *Merchants of Doubt: How a Handful of Scientists Obscured the Truth on Issues from Tobacco Smoke to Global Warming* (London: Bloomsbury, 2010).

18 Clare Roepke and Eric Schaff. 'Long Tail Strings: Impact of the Dalkon Shield 40 Years Later.' *Open Journal of Obstetrics and Gynecology* 4, no. 16 (2014): 996–1005. doi: 10.4236/ojog.2014.416140;

Elizabeth Siegel Watkins, 'From Breakthrough to Bust: The Brief Life of Norplant, the Contraceptive Implant.' *Journal of Women's* History 22, no. 3 (2010): 88–111. doi:10.1353/jowh.2010.0585.

19 Vostral, *Toxic Shock*.

20 Roy Porter, *The Greatest Benefit to Mankind: A Medical History of Humanity* (New York: W.W. Norton, 1999).

21 An example is the phenomenon of shoplifting by upper-class White women being named *kleptomania* – the obsessive desire to steal goods from the new commercial spaces of department stores – and this disease diagnosis shaped by gender, race, and class. See Elaine Abelson, 'The Invention of Kleptomania.' *Signs* 15, no. 1 (1989): 123–43.

22 Barbara Ehrenreich and Deirdre English, *For Her Own Good: 150 Years of Experts' Advice to Women* (New York: Doubleday, 1978); Judith Walzer Leavitt (ed.), *Women and Health in America: Historical Readings*, 2nd edition (Madison, WI: University of Wisconsin Press, 1999); Rima Apple (ed.), *Women, Health and Medicine in America: A Historical Handbook* (New Brunswick, NJ: Rutgers University Press, 1992).

23 Judith Walzer Leavitt, *Brought to Bed: Childbearing in America, 1750–1950* (New York: Oxford University Press, 1986).

24 Vostral, *Under Wraps*; Freidenfelds, *The Modern Period*.

25 Vostral, *Under Wraps*; Laura Briggs, 'The Race of Hysteria: "Overcivilization" and the "Savage" Woman in Late Nineteenth-Century Obstetrics and Gynecology.' *American Quarterly* 52, no. 2 (2000): 246–73.

26 See Brumberg, *The Body Project*; however, she does not use the framing of technology to make this point.

27 Freidenfelds, *The Modern Period*.

28 Susan Strasser, *Waste and Want: A Social History of Trash* (New York: MacMillan, 1999).

29 Danya Glabau, 'Food Allergies and the Hygienic Sublime.' *Catalyst: Feminism, Theory, Technoscience* 5, no. 2 (2019): 1–26. On transforming the body, see also Rebecca Herzig, *Plucked: A History of Hair Removal* (New York: New York University Press, 2015).

30 Lisa Rosner (ed.), *The Technological Fix: How People Use Technology to Create and Solve Problems* (New York: Routledge, 2004).

31 Anon., 'Follow-Up on Toxic-Shock Syndrome – United States.' *MMWR* 29, no. 25 (27 June 1980): 297–9; Anon., 'Follow-Up on Toxic-Shock Syndrome.' *MMWR* 29, no. 37 (19 September 1980): 441–5.

32 Vostral, *Toxic Shock*.

33 Tom Riley, *The Price of Life: One Woman's Death from Toxic Shock* (Bethesda, MD: Adler & Adler, 1986).

34 *Michael L. Kehm v. Procter & Gamble.* United States Courthouse, Cedar Rapids, Iowa. 5 April 1982: p. 2568.

35 While given names are part of the public court records, in order to maintain anonymity, they have been removed here.

36 *Kehm v. Procter & Gamble*, pp. 2650–2.

37 Anon., 'Toxic Shock Syndrome – United States, 1970–1980.' *MMWR* 30, no. 3 (30 January 1981): 25–36; Anon., 'Update: Toxic Shock Syndrome – United States.' *MMWR* 32, no. 30 (5 August 1983): 398–400.

38 The premise of the women's health movement begun during the late 1960s and 1970s challenged this type of thinking, though the rationale provided by Procter & Gamble was indicative of the dismissal of women's health. See Wendy Kline, *Bodies of Knowledge: Sexuality, Reproduction, and Women's Health in the Second Wave* (Chicago, IL: University of Chicago Press, 2010); Jennifer Nelson, *More than Medicine: A History of the Feminist Women's Health Movement* (New York: NYU Press, 2015).

39 Anon., 'Toxic Shock Case Weighed by Court.' *New York Times*, 4 November 1982.

40 Gene Matthews, 'Toxic Shock Syndrome: A Lasting Legacy', CDC 'We Were There' Lecture Series, accessed October 10, 2023 19 October 2017. www.cdc.gov/od/science/wewerethere/toxicshock/index.html. Accessed 10 October 2023.

41 Matthews, 'Toxic Shock Syndrome'.

42 Matthews, 'Toxic Shock Syndrome'; Anon., 'Epidemiologic Notes and Reports Toxic-Shock Syndrome – United States.' *MMWR* 46, no. 22 (6 June 1997): 492–5.

43 Matthews, 'Toxic Shock Syndrome'.

44 *Lampshire v. Procter & Gamble*, 94 F.R.D 58 (N.D. Ga 1982) appealed; Lampshire District Court decision vacated by 11th Circuit of procedural grounds, 708 F.2d 732 (11th Cir 1983) vacated and settled; *Farnsworth v. P&G v. CDC*, 101 F.R.D. 355 (N.D. Ga 1984) – won and Procter & Gamble appealed; *Farnsworth v. P&G*, 758 F.2d 1545 (11th Cir. 1985).

45 *Farnsworth v. P&G v. CDC*, 758 F.2d 1545 (11th Cir. 1985). https://law.resource.org/pub/us/case/reporter/F2/758/758.F2d.1545.84–8330.html. Accessed 10 October 2023.

46 *Farnsworth v. P&G v. CDC.*

47 Eden Osucha, 'The Whiteness of Privacy: Race, Media, Law.' *Camera Obscura: Feminism, Culture, and Media Studies* 24, no. 170 (2009): 67–107.

48 *Farnsworth v. P&G v. CDC.*

49 Matthews, 'Toxic Shock Syndrome'.
50 *Farnsworth v. P&G v. CDC.*
51 Curran, 'Protecting Confidentiality'.
52 Curran, 'Protecting Confidentiality', 1028.
53 Robert Baker and Linda Emanuel, 'The Efficacy of Professional Ethics: The AMA Code of Ethics in Historical and Current Perspective.' *The Hastings Center Report*, 30, no. 4 (July–August 2000): S13–S17.
54 Elizabeth Siegel Watkins, *On the Pill: A Social History of Oral Contraceptives* (Baltimore, MD: Johns Hopkins University Press, 1998); Leslie Reagan, *When Abortion Was a Crime: Women, Medicine, and Law in the United States, 1867–1973* (Oakland, CA: University of California Press, 1997); Linda Gordon, *The Moral Property of Women: The History of Birth Control Politics in America* (Champaign, IL: University of Illinois Press, 2007).
55 Heidi Schreck, Interview with Terri Gross. 'How Women Have Been "Profoundly" Left Out of the U.S Constitution.' *Fresh Air*, 20 March 2019. www.npr.org/2019/03/20/705077773/how-women-have-been-profoundly-left-out-of-the-constitution. Accessed 10 October 2023.
56 Virginia Eubanks, 'Want to Predict the Future of Surveillance? Ask Poor Communities.' *The American Prospect*, 15 January 2014. https://prospect.org/power/want-predict-future-surveillance-ask-poor-communities/. Accessed 10 October 2023.
57 Danielle Whicher, Mahnoor Ahmed, Sameer Siddiqi, Inez Adams, Maryan Zirkle, Claudia Grossmann, and Kristin L. Carman (eds), 'Health Data Sharing to Support Better Outcomes: Building a Foundation of Stakeholder Trust.' National Academy of Medicine, 2021. https://nam.edu/health-data-sharing-special-publication/. Accessed 10 October 2023

Bibliography

Abelson, Elaine. 'The Invention of Kleptomania.' *Signs* 15, no. 1 (1989): 123–43.
Anonymous. 'Epidemiologic Notes and Reports Toxic-Shock Syndrome – United States.' *MMWR* 46, no. 22 (6 June 1997): 492–5.
Anonymous. 'Follow-Up on Toxic-Shock Syndrome – United States.' *MMWR* 29, no. 25 (27 June 1980): 297–9.
Anonymous. 'Follow-Up on Toxic-Shock Syndrome.' *MMWR* 29, no. 37 (19 September 1980): 441–5.
Anonymous. 'Toxic Shock Case Weighed by Court.' *New York Times*, 4 November 1982.
Anonymous. 'Toxic Shock Syndrome – United States, 1970–1980.' *MMWR* 30, no. 3 (30 January 1981): 25–36.

Anonymous. 'Update: Toxic Shock Syndrome – United States.' *MMWR* 32, no. 30 (5 August 1983): 398–400.

Apple, Rima (ed.). *Women, Health and Medicine in America: A Historical Handbook.* New Brunswick, NJ: Rutgers University Press, 1992.

Baker, Robert, and Linda Emanuel. 'The Efficacy of Professional Ethics: The AMA Code of Ethics in Historical and Current Perspective.' *The Hastings Center Report* 30, no. 4 (July–August 2000): S13–S17.

Brandt, Allan. *The Cigarette Century: The Rise, Fall, and Deadly Persistence of the Product That Defined America.* New York: Perseus Books Group, 2007.

Briggs, Laura. 'The Race of Hysteria: "Overcivilization" and the "Savage" Woman in Late Nineteenth-Century Obstetrics and Gynecology.' *American Quarterly* 52, no. 2 (2000): 246–73.

Brown, Phil. *Toxic Exposures: Contested Illnesses and the Environmental Health* Movement. New York: Columbia University Press, 2007.

Brumberg, Joan Jacobs. *The Body Project: An Intimate History of American Girls.* New York: Vintage Books, 1998.

Curran, William. 'Protecting Confidentiality in Epidemiologic Investigations by the Centers for Disease Control.' *New England Journal of Medicine* 314, no. 16 (17 April 1986): 1027–8.

Davis, Jeffrey. 'The Investigation of TSS in Wisconsin and Beyond, 1979–1980', in *Outbreak Investigation around the World: Case Studies in Infectious Disease Field Epidemiology*, edited by Mark Dworkin, pp. 79–201. Boston, MA: Jones and Bartlett, 2009.

Ehrenreich, Barbara, and Deirdre English. *For Her Own Good: 150 Years of Experts' Advice to Women.* New York: Doubleday, 1978.

Eubanks, Virginia. 'Want to Predict the Future of Surveillance? Ask Poor Communities.' *The American Prospect*, 15 January 2014. https://prosp ect.org/power/want-predict-future-surveillance-ask-poor-communities/. Accessed 10 October 2023.

Farnsworth v. P&G v. CDC, 758 F.2d 1545 (11th Cir. 1985). https://law. resource.org/pub/us/case/reporter/F2/758/758.F2d.1545.84–8330.html. Accessed 10 October 2023.

Farnsworth v. Procter & Gamble Company v. Center for Disease Control, 758 F.2d 1545 (11th Cir. 1985). https://law.resource.org/pub/ us/case/reporter/F2/758/758.F2d.1545.84–8330.html. Accessed 10 October 2023.

Freidenfelds, Lara. *The Modern Period: Menstruation in Twentieth-Century America.* Baltimore, MD: Johns Hopkins University Press, 2009.

Glabau, Danya. 'Food Allergies and the Hygienic Sublime.' *Catalyst: Feminism, Theory, Technoscience* 5, no. 2 (2019): 1–26.

Gordon, Linda. *The Moral Property of Women: The History of Birth Control Politics in America.* Champaign, IL: University of Illinois Press, 2007.

Herzig, Rebecca. *Plucked: A History of Hair Removal.* New York: New York University Press, 2015.

Jain, Sarah Lochlann. *Injury: The Politics of Product Design and Safety Law in the United States*. Princeton, NJ: Princeton University Press, 2006.

Johnson, Margaret. 'The Period Is Political: Menstrual Justice, Abortion Rights, and Reproductive Justice.' Keynote address for the bi-annual meeting of the Society for Menstrual Cycle Research, Washington DC, 2023.

Kline, Wendy. *Bodies of Knowledge: Sexuality, Reproduction, and Women's Health in the Second Wave*. Chicago, IL: University of Chicago Press, 2010.

Lampshire v. Procter & Gamble, 94 F.R.D 58 (N.D. Ga 1982).

Leavitt, Judith Walzer. *Brought to Bed: Childbearing in America, 1750–1950*. New York: Oxford University Press, 1986.

Leavitt, Judith Walzer (ed.). *Women and Health in America: Historical Readings*, 2nd edition. Madison, WI: University of Wisconsin Press, 1999.

Matthews, Gene. 'Toxic Shock Syndrome: A Lasting Legacy.' CDC 'We Were There' Lecture Series, 19 October 2017. www.cdc.gov/od/science/wewerethere/toxicshock/index.html. Accessed 10 October 2023.

Michael L. Kehm v. Procter & Gamble. United States Courthouse, Cedar Rapids, Iowa. 5 April 1982: p. 2568.

Mold, Alex. *Making the Patient-Consumer: Patient Organisations and Health Consumerism in Britain*. Manchester: Manchester University Press, 2015.

Nelson, Jennifer. *More than Medicine: A History of the Feminist Women's Health Movement*. New York: NYU Press, 2015.

Oreskes, Naomi, and Erik M. Conway. *Merchants of Doubt: How a Handful of Scientists Obscured the Truth on Issues from Tobacco Smoke to Global Warming*. London: Bloomsbury Press, 2010.

Osucha, Eden. 'The Whiteness of Privacy: Race, Media, Law.' *Camera Obscura: Feminism, Culture, and Media Studies* 24, no. 170 (2009): 67–107.

Porter, Roy. *The Greatest Benefit to Mankind: A Medical History of Humanity*. New York: W.W. Norton, 1999.

Reagan, Leslie. *When Abortion Was a Crime: Women, Medicine, and Law in the United States, 1867–1973*. Oakland, CA: University of California Press, 1997.

Riley, Denise. *Am I that Name? Feminism and the Category of 'Women' in History*. London: Macmillan, 1988.

Riley, Tom. *The Price of Life: One Woman's Death from Toxic Shock*. Bethesda, MD: Adler & Adler, 1986.

Roepke, Clare, and Eric Schaff. 'Long Tail Strings: Impact of the Dalkon Shield 40 Years Later.' *Open Journal of Obstetrics and Gynecology* 4, no. 16 (2014): 996–1005. doi: 10.4236/ojog.2014.416140.

Rosner, Lisa (ed.). *The Technological Fix: How People Use Technology to Create and Solve Problems*. New York: Routledge, 2004.

Røstvik, Camilla Mørk. 'Crimson Waves: Narratives about Menstruation, Water, and Cleanliness.' *Visual Culture and Gender* 13 (2018): 54–63. http://vcg.emitto.net/index.php/vcg/article/view/114/118. Accessed 10 October 2023.

Schreck, Heidi. Interview with Terri Gross. 'How Women Have Been "Profoundly" Left Out of the U.S. Constitution.' *Fresh Air*, 20 March 2019. www.npr.org/2019/03/20/705077773/how-women-have-been-profoundly-left-out-of-the-constitution. Accessed 10 October 2023.

Strasser, Susan. *Waste and Want: A Social History of Trash*. New York: MacMillan, 1999.

Tierno, Philip. *The Secret Life of Germs: Observations and Lessons from a Microbe Hunter*. New York: Pocket Books, 2004.

Tomes, Nancy. *Remaking the American Patient: How Madison Avenue and Modern Medicine Turned Patients into Consumers*. Chapel Hill, NC: University of North Carolina Press, 2016.

Vostral, Sharra. *Toxic Shock: A Social History*. New York: New York University Press, 2018.

Vostral, Sharra. *Under Wraps: A History of Menstrual Hygiene Technology*. Lanham, MD: Lexington Press, 2008.

Watkins, Elizabeth Siegel. 'From Breakthrough to Bust: The Brief Life of Norplant, the Contraceptive Implant.' *Journal of Women's History* 22, no. 3 (2010): 88–111. doi:10.1353/jowh.2010.0585.

Watkins, Elizabeth Siegel. *On the Pill: A Social History of Oral Contraceptives*. Baltimore, MD: Johns Hopkins University Press, 1998.

Whicher, Danielle, Mahnoor Ahmed, Sameer Siddiqi, Inez Adams, Maryan Zirkle, Claudia Grossmann, and Kristin L. Carman (eds). 'Health Data Sharing to Support Better Outcomes: Building a Foundation of Stakeholder Trust.' National Academy of Medicine, 2021. https://nam.edu/health-data-sharing-special-publication/. Accessed 10 October 2023.

5

Just stories: Side effects and the patient voice online

Antoine Lentacker

Do antidepressants cause alcohol dependence?

A hospital employee in the south of England, Anne-Marie, had been prescribed paroxetine in her mid-thirties to help alleviate an eating disorder she developed following the sudden death of her father. The drug seemed to work as intended. Her fear of choking subsided within a few weeks and she was able to eat normally again. Meanwhile, several of her friends started remarking on her increased drinking. In denial at first, she lost contact with friends and family as her troubles piled up. She suffered car crashes and was arrested several times for compulsive nuisance calls she made to her local police station when under the influence. Seeing her life unravel, she turned to the internet in search for answers and stumbled on reports from other people who complained of intense cravings for alcohol while taking selective serotonin reuptake inhibitors, or SSRIs, the class of antidepressants to which paroxetine belonged. Those reports, she later wrote, 'shocked' her.

Sensing she had found the key to her issues, she printed out some of those reports and brought them to her doctor. 'Sympathetic but not convinced', as she described, he agreed to change her prescription in 2005 but switched her to citalopram, another member of the SSRI class, only to see her troubles worsen. She lost her job, spent time in prison, went through rehab programmes, but could not overcome her self-destructive cravings. She also stopped talking to doctors about the possible link between the drugs and her alcoholism. Denial, they explained, was a symptom of her drinking problem. Drugs like paroxetine or citalopram were often given to help alcoholics in their recoveries; they couldn't be causing a problem they were meant to treat. Perhaps the doctors were right?

Yet Anne-Marie didn't give up. Scattered across depression forums and websites like the International Coalition for Drug Awareness, the Seroxat Users Support Group, or Seroxat Secrets (Seroxat was paroxetine's brand name in the United Kingdom), she continued to read accounts from people whose experiences seemed to mirror her own. Her search for academic studies on SSRIs and alcohol dependence yielded little. Nothing seemed to have been published on the subject, until a post on one forum mentioned a 1994 Yale study that explored the link between alcohol use and the serotonin system. Knowing the drugs she had been given acted on serotonin levels, she dug up further research papers on the serotonin system and slowly pieced together an explanation of how her drug might be linked to her drinking problem. Despite debilitating withdrawal symptoms, she forced herself off citalopram and her cravings began to diminish. She further discovered that not all antidepressants act on the serotonin system and in 2010 persuaded her doctor to prescribe mirtazapine, an atypical antidepressant that acts as a serotonin antagonist. As soon as she started her new medication, her by then decade-old drinking problem fully cleared up.[1]

Stories like this one epitomize the promise of the Web as a technology to harness patient knowledge and reinvent drug safety. Since it opened to the public in the mid-1990s, the World Wide Web has connected people affected by similar health conditions and moved by a same desire to 'get to the bottom of it', as Anne-Marie put it. As an unprecedented mass of information on diseases and therapies – including the knowledge once tucked away between the covers of printed journals – became accessible online, patients saw opportunities to take on more active roles in the work of diagnosing and managing their illnesses. Those new possibilities eventually enabled Anne-Marie, in a somewhat awkward partnership with her physician, to overcome the issue that had dogged her for so many years. At the same time, however, her story could be taken as evidence of how little that promise has been realized. The healthcare professionals she interacted with met her efforts to make sense of her condition with scepticism or resistance. Her insights were not heeded but rather dismissed as symptoms of a mental disorder. Nor was her experience unusual. Doctors' inability to listen to their patients seemed to be a recurring theme on the online forums she frequented. Two decades after the advent of the Web, there was still no obvious place in the therapeutic process for the sort of crowd-sourced patient knowledge she sought to volunteer.

This chapter locates the uncertainties about the role of 'the patient's voice' in the inherent ambiguities of the concept of voice itself. The at once unstable and overdetermined nature of voice as a concept has been explored by others. Derrida, most famously, identified the notion of voice – of knowledge embodied, intuitive and unmediated, as opposed to the dead letter of written speech, which is no longer genuine knowledge unless it is brought back to life in the inner or outward voice of a conscience – as a foundational belief of metaphysics. But the concept of voice as living knowledge that underwrites what Derrida called 'logocentrism' is itself a metaphor. Knowledge is recognition; it is knowing something as something else. As such, there is no unmediated knowledge, purely present and transparent to itself. Voice becomes knowledge when it is articulated, differentiated, and in some ways alienated in a sign, a trace whose meaning can be reactivated. There is no escape from the vagaries of representation.[2]

The patient's voice, therefore, ought to be analysed as something that acquires meaning and effect only in and through certain representational technologies. The emergence of the Web in the 1990s initially revived a certain metaphysics of voice, a celebration of unmediated expression and a rejection of the established division of labour between producers and consumers of knowledge. In medicine, as in other domains, the Web promised to liberate lay voices from the grip of experts or gatekeepers. Starting in the late 1990s, surveys and studies highlighted how quickly patients embraced the Web for both information and community, as discussion boards, blogs, and other forums carried patients' words beyond the isolating here and now of illness or disability, disrupted the top-down dispensation of medical expertise, and reshaped the very experience of disease.[3] Yet far from being a pure and transparent medium for the patient's voice, the Web is itself a complex representational infrastructure that amplifies the voices of users by alienating them in a trace that can be reproduced and disseminated. Transmuted into digital object, typically in written form, it is detached from the sick body, deferred and displaced, out of time and out of place. This severing of the voice from its living source who, as in the case of

Anne-Marie, often hides behind anonymity is also what makes the Web suspect as a source of knowledge. While suspicion towards medical information online has long prevailed among physicians, it has spread more widely in recent years, particularly since the beginning of the COVID-19 pandemic. For a range of reasons – the main one being the gradual transformation of the internet from a technology of free information exchange into an instrument of monopoly capitalism based on new forms of commercial surveillance and consumption – a darker view of the Web as a medium of misinformation, a site for the proliferation of unverified accounts, is taking the place of the logocentric utopias of the early years.[4] Hence, one could say of the online voice what Derrida, commenting on a metaphor by Plato, said of writing: it is a pharmakon, both remedy and poison, a supplement that is at once supporting and subverting, extending and extinguishing the voice of the living and knowing subject.

To trace the effects of this paradox over the last thirty years, I examine the history of one website whose trajectory intersected with Anne-Marie's. RxISK.org, as it is called, was created by a trio of scholars – David Healy, a psychiatrist and author in the United Kingdom; Dee Mangin, a professor of Family Medicine in New Zealand; and Kalman Applbaum, an anthropologist in the United States – to uncover side effects by soliciting adverse drug reaction reports directly from those who take (rather than those who prescribe) pharmaceutical drugs. While RxISK.org came into existence in 2012, its roots lie in a longer struggle to bring to light the unrecognized side effects of SSRIs, a struggle that coincided in large part with the advent of the Web but also with the rise of 'evidence-based medicine'. This history, I show, sets it apart from other digital health platforms that seek to harness patient voices, both in its forms of expertise as well as its data collection practices. RxISK.org's distinctive place in the landscape of eHealth makes it a uniquely revealing vantage point to reflect on the conditions under which the patient consumer may also become a genuine producer of pharmaceutical knowledge.[5]

How to hide a side effect

Controversies around the side effects of SSRI antidepressants threw into sharp relief this tension between experience and expertise. The first commercially successful SSRI, Eli Lilly's Prozac (fluoxetine), came on to the United States market in 1988. Unlike older generations of antidepressants, which dated back to the 1950s and 1960s, it had undergone extensive clinical trials before its introduction. Those trials, according to the company, showed the drug to be as effective as its older counterparts but better tolerated and far safer in overdose. The drug's appeal derived in large part from its allegedly favourable side effect profile. Psychiatrists, soon followed by primary care physicians, prescribed it for a widening range of complaints. Tapping into a vast market for previously undiagnosed or unmedicated mood disorders, Lilly's drug quickly generated over a billion dollars in yearly sales.[6] Yet the company's assurances about the safety of its new product did not go unchallenged for long. Starting in 1990, several reports appeared in medical journals describing cases of patients that appeared to develop intense suicidal preoccupations within days of initiating treatment with the novel molecule. Reports about the drug's 'suicide side effect' cast a shadow over its prospects, even more so given the unprecedented amount of attention Prozac garnered in the general media. Lilly moved to defend its product, sparking an extended controversy in which the emerging medium of the Web would come to play an important role.

Initially, the arguments on both sides were given voice in a rather traditional forum. Following the first reports of Prozac's possible link to suicidal or violent behaviour, the United States consumer organization Public Citizen petitioned the Food & Drug Administration (FDA) to add on Prozac's label a warning about this previously unrecognized effect. In September 1991, as it readied itself to approve Pfizer's Zoloft (sertraline) and SmithKline's Paxil (paroxetine), the next bestselling members in the expanding SSRI family, the FDA convened a committee of experts to assess the evidence on the link between antidepressants and suicide. The day-long hearing opened with testimonies from the public. About two dozen speakers related intensely personal tragedies of which Prozac, they claimed, had been the cause. Several survivors told their own stories, while relatives spoke for those who had not survived. The accounts they gave ranged from cases of suicides completed within days of going on the drug to cases of debilitating withdrawal symptoms experienced after months of treatment. Two testimonies described homicides, and a further one a homicide attempt. Multiple witnesses noted how the erratic behaviour they or their relatives exhibited on Prozac seemed entirely 'out of character', or bore no resemblance to the mood disorders they may have experienced before their encounter with the drug. Some speakers explicitly invoked their authority as survivors, as 'living proof' of 'silenced adverse effects'.[7]

In contrast to the open hearing of the morning, which took place in an atmosphere saturated with drama, the afternoon sessions proceeded behind closed doors and consisted entirely of hard-headed discussions among the outside experts, FDA staff, and Lilly's representatives. Every one of the speakers opened their presentation with an acknowledgement of the power of the personal accounts delivered during the morning. Dr Nina Schooler found herself 'enormously touched and responsive to a number of the stories we heard this morning', many of which came from individuals with no prior history of suicidality. Paul Leber from the FDA acknowledged that those stories tended to echo the reports published in the literature as well as the nearly 900 adverse event reports that the agency had received regarding Prozac and its connection to violent or suicidal behaviour. Yet those admissions were all punctuated with the observation that stories remain anecdotes unless they are backed by controlled studies. Taken together, the reports were described as a 'numerator without denominator', as a 'picture' but only a 'preliminary' one, as 'troubling' and 'hard to ignore', but also 'difficult to interpret'. By the early 1990s, the years in which the discourse of evidence-based medicine took shape, the inscrutability of single cases and the sense that self-knowledge fades all too easily into self-deception had become basic tenets of biomedicine's epistemology. The experts who spoke on Lilly's behalf based their case entirely on those general principles. Only randomized trials in which both patients and clinicians are blinded and single cases are controlled, they argued, can decide the question of causality in a scientific manner. In Lilly's trial database, suicides were no more frequent among patients who had been given the drug than among those given a placebo. In the end, the panel's vote broke comfortably in favour of the company and the FDA declined to issue the warnings requested by Public Citizen.

The same strategic embrace of evidence hierarchies that hailed randomized trials as the 'gold standard' of evidence and devalued individual cases as mere anecdotes served the industry in the legal arena as well. Following the FDA's ruling, most lawsuits filed by plaintiffs who had lost family members to a suicide or act of violence involving Prozac settled on terms favourable to Lilly. A small handful of cases, however, came before a jury. Wary of the power that stories from sympathetic victims might exercise over lay jurors, company lawyers argued in court that no drugs in the history of psychiatry had been as thoroughly researched as SSRIs, that the trial data on them had been scrutinized by the FDA, analysed and meta-analysed in the medical literature, and were found to contain no credible evidence of a link between the drugs and the harms for which they were blamed. In this context, no amount of case-specific evidence, the sort of first-hand evidence provided by plaintiffs on the stand, should be considered relevant to the outcome of a trial.[8] For about a decade, this defence worked as companies intended. An expansive concept of 'anecdotal' evidence that lumped single cases, however well documented, with other varieties of hearsay, of unverified or unreliable evidence, helped strip the voices of patients, relatives, or treating physicians of their probative force, thus ensuring that cases against SSRI manufacturers were either settled, dismissed, or defeated in court.

Anecdote and evidence

Despite the setbacks, the work of litigation laid the ground for new ways of thinking about drug safety. The legal discovery process, which empowers parties in a lawsuit to obtain relevant evidence from each other, gave plaintiffs' attorneys access to reams of internal company documents. These records made clear that SSRI manufacturers had long been aware of the triggering effect that the molecules seemed to have on some patients and had taken steps to disguise this effect in clinical trial data. These included the screening out of trial participants with prior histories of suicidality, the loose or misleading coding of adverse events, the ghostwriting of trial reports, and the selective publication of data. Serving as expert witness in several SSRI cases, the Irish psychiatrist David Healy worked with plaintiffs and their attorneys to reveal how such practices distorted the scientific record in a way that highlighted benefits and concealed harms, years before the design and conduct of clinical trials attracted broader scholarly attention. This experience formed the basis of the later critique of the drug industry for which Healy is now best known – and an impetus also of the RxISK project he eventually undertook with Dee Mangin and Kalman Applbaum.[9]

Healy's work on behalf of plaintiffs went beyond exposing the distortion of the evidence base generated by corporate clinical trials. He also argued for a rehabilitation of forensic methods in the adjudication of drug-related injury. As he testified in court, Austin Bradford Hill, the English statistician who introduced randomized trials in drug testing, never regarded controlled trials as the sole or even the best way to determine causality. In 1965 Bradford Hill had outlined nine factors to consider in making such determinations.[10] Some of these criteria – for example, 'temporality' (the effect must occur within a stable timeframe after consumption of the drug); 'plausibility' (the assumed cause-and-effect connection should be biologically plausible, ideally more so than alternative explanations); or 'consistency' (the effect is analogous to outcomes observed in other patients, by other clinicians, or with other drugs) – could be evaluated in single cases. One way of doing so in side effect research was through so-called challenge–dechallenge–rechallenge protocols, whereby a patient who exhibits a certain reaction upon taking a drug is taken off the drug and then exposed to it again in order to verify if a similar reaction reoccurs within a similar timeframe. Such experiments, Healy observed, are also controlled studies, though the controlling is done across time in the same individual rather than across a pool of different individuals randomly assigned to different arms of a trial.[11]

One of the cases in which Healy testified, *Tobin v. SmithKline Beecham*, tried in Wyoming in 2001, happened to involve a patient named Don Schell who had experienced spells of stress and depression at various points in his professional life. Shortly after Prozac came on the market, his doctor prescribed him the new drug alongside some anti-anxiety pills to help him through one such spell. The anxiolytic medication notwithstanding, Schell became intensely agitated and stopped the treatment. When he suffered a renewed bout of depression in 1998, nearly ten years later, a hospital resident handed him a Paxil sample, though this time without a covering prescription for anxiety medication. The resident was unaware of Schell's prior reaction to Prozac, and Schell was unaware that

Paxil, also a member of the SSRI class, works in essentially the same ways as Prozac. Forty-eight hours later, he fatally shot his wife, daughter, infant granddaughter, and himself. The grisly sequence of events turned Schell's history into a natural challenge–dechallenge–rechallenge experiment, which Healy argued provided stronger evidence of causation than a clinical trial could ever be expected to do. The jury was persuaded, and in June 2001 Schell's son-in-law, Tim Tobin, became the first plaintiff ever to win a verdict against a drug company for a psychiatric side effect of one of its products. The rare legal victory highlighted the role that fact-finding practices dedicated to the particulars of a single case could play in understanding drug-related injury.

Still, it was the broadcasting of patient voices that compelled the eventual recognition of the reality of SSRI-induced suicides. In October 2002 the BBC dedicated an episode of its popular public affairs programme *Panorama* to the side effects of Paxil, which in the United Kingdom was sold under the brand name Seroxat and was the country's most widely prescribed antidepressant. *Secrets of Seroxat* told the story of the Tobin trial but consisted primarily of interviews with patients who spoke before the camera about their personal struggles with the drug, ranging from debilitating withdrawal syndromes to self-harm and suicide attempts. Seen by an estimated four million viewers, the programme elicited a public response unprecedented in the history of BBC TV: 65,000 callers and a further 1,400 or so email writers reached the channel, most of them to share their own stories with SmithKline's drug. Clearly, the extent of the hidden suffering caused by the drug exceeded anyone's suspicions. The groundswell of public attention suddenly brought on to those harms moved the United Kingdom's drug regulator, the Medicines and Healthcare Products Regulatory Agency (MHRA), to revisit the evidence on the safety of SSRIs. Internal company data disclosed to the MHRA during the ensuing review directly contradicted the reassuring findings about the drug's safety published in the medical literature. In particular, the trials sponsored by SmithKline Beecham (which had become GSK in the meantime) to license Paxil/Seroxat for paediatric mood disorders had shown no efficacy in treating depression but appeared to double the risk of suicide or self-harm among adolescents. Those findings notwithstanding, the company commissioned the preparation and publication of reports based on truncated data that declared the drugs safe and effective for use in patients under eighteen. When those actions came to light in the spring of 2003, reluctant regulators were forced at long last to warn doctors and patients of the reality of SSRI-induced suicides.[12]

Seen in hindsight, the decade-long struggle to bring the hazards of SSRIs to light taught a twofold lesson about the regulatory arrangements relied upon to guarantee the safety of prescription drugs. The first concerned the limits of clinical trials with respect to identifying uncommon or unforeseen drug reactions, especially when those trials are commissioned by the industry and their findings are owned by the companies that pay for them. A rich literature in both medicine and the social sciences has since outlined the nature of those limits and shown how the industry exploits them to suppress evidence about the hazards of drugs.[13] Subsequent revelations about the hidden harms of other bestselling drugs such as Merck's arthritis drug Vioxx in 2004 or Lilly's antipsychotic Zyprexa in 2006 suggested that SSRIs' misrecognized link to suicide was no isolated accident.[14] Rather, it reflected systemic features of the contemporary economy of pharmaceutical knowledge. The inevitable outcome of grounding clinical guidelines in a distorted evidence base that highlights benefits and downplays harms is, in Dee Mangin's words, an 'invisible epidemic of iatrogenic illness'.[15] In this context, the careful parsing of case reports of adverse drug reactions appeared all the more critical, particularly when they suggest adverse reactions that, according to clinical trial findings, were not supposed to occur.

The second lesson, then, concerned the importance of obtaining side effect reports directly from patients. Following the broadcasting of *Secrets of Seroxat*, BBC journalists Shelley Jofre and Andrew Bell partnered with drug safety experts Charles Medawar and Andrew Herxheimer so they could review the 1,400 or so emails they received from viewers about Seroxat and its hazards. Medawar was the director of the London-based Public Interest Research Centre, an offshoot of Public Citizen in the United Kingdom. His own work had focused since the 1980s on issues of prescription drug policy. Equipped with a computer, a '128k modem', and an early 'strategic sense' of the internet's potential as a tool to 'exchange information and consolidate experiences', as he put it in a 2002 interview with Healy, Medawar began scrutinizing some of the early message boards to which antidepressant users turned to share their experiences.[16] In 1997 he published 'the Antidepressant Web', a study that drew in part on material collected on online forums, and set up a website – socialaudit.org.uk – to disseminate his research findings and host discussion boards specifically dedicated to the unacknowledged harms of SSRIs.[17] Herxheimer was a pharmacologist with a distinguished career in London and Oxford as well as a dedicated patient advocate. Focusing on self-harm and withdrawal reactions, the two harms highlighted in *Secrets of Seroxat*, Herxheimer and Medawar conducted an in-depth comparison of how side effects were described by users in their emails to the BBC and by doctors in adverse drug event (ADE) reports filed with the MHRA.

In the United Kingdom, as in most other countries at the time, drug authorities solicited ADE reports exclusively from health professionals. Pharmacovigilance relied on physician reporting, not patient reporting, of side effects. 'We believe the underlying reason for regulators' disdain', Herxheimer and Medawar wrote, 'is their prejudice that what a patient reports is "anecdotal" and does not constitute "scientific" evidence, and therefore should not be accepted without confirmation by a professional'.[18] And, indeed, they acknowledged that the emails formed a wholly unsystematic dataset. They came from a self-selecting group of viewers with overwhelmingly negative experiences on GSK's drug. The absence of random sampling and of a control group made it unsuitable for any estimation of ratios and frequencies. Many emails lacked key information such as the user's age, sex, dosage, or diagnosis. They were quintessentially anecdotal in the sense of evidence-based medicine. And yet the authors found them superior in many respects to the reports supplied by physicians. Under the delegated system favoured by drug regulators, they observed: 'the patient's report is filtered through the doctor's own expectations and his or her interpretation of what is credible, serious, relevant, or worth reporting'. As 'translation[s] in medical shorthand of what the patient says', doctors' reports are terse (averaging forty–seventy-five words per report) and stripped of any textured description of the lived experience of side effects and of their consequences on relationships, employment, or mobility.[19] Those aspects, by contrast, were remarkably salient in the first-person accounts contained in the emails. Citing those accounts at length, Medawar and Herxheimer affirmed 'the value of "immersion"' in a rich, albeit haphazardly assembled, collection of narrativized accounts.[20] Apprehended holistically, self-narratives yielded a pregnant picture not only of the reality of these side effects but also of their underappreciated impact on the quality of life of those who experienced them – 'what withdrawal problems, weight gain, suicidal behaviour, or loss of libido actually mean in the context of personal and social life'.[21]

How to reveal a side effect

In the mid-2000s, the need to rethink the place of the patient's voice in the therapeutic process and the possibilities that digital communication technologies afforded in that respect became a topic of broader conversation. In the wake of the revelations about SSRIs' and Vioxx's hidden harms, the adverse drug reaction (ADR) reporting systems of most drug agencies in Europe and North America began accepting submissions from patients. In nearly all countries, it also became possible for patients and providers to submit their reports directly online.[22] More importantly, changes in the economy of the Web also transformed the uses and meanings of the patient voice online. At the time, those changes were subsumed under the banner of Web 2.0 and described as a shift towards interactive platforms and user-generated content. More recent and critical analyses, however, have linked the technological shifts behind 'Web 2.0' platforms to the rise of surveillance capitalism, a new accumulation regime based on the extraction of personal data and their conversion into a new kind of capital. As scholars have shown, some of the most prominent 'eHealth' platforms, such as PatientsLikeMe, fit squarely within this new digital economy. They began appearing in the mid-2000s and have since exploited the rhetoric of openness, participation, and empowerment on an entirely new scale to obtain from users the data they subsequently monetize.[23]

The RxISK project fitted in this moment as well, a moment in which the internet felt, in Healy's words, like 'a tool that might break things open, might be a force for consumers to get their voices out'.[24] But the project's 'intertwined intellectual, political, and practical … roots', to borrow Applbaum's words, diverged from those of commercial eHealth sites.[25] RxISK's co-creators met as participants in a global and cross-disciplinary conversation that took shape amid revelations regarding the hidden harms of SSRIs and other blockbuster medications. The 'central purpose' of their project was to 'get around the firewall of corporate-owned clinical trial data' and to 'generate and bring uncorrupted, which is to say non-corporate owned, data to bear upon public knowledge of the drugs we take'.[26] Those goals were fundamentally at odds with those of the main commercial eHealth sites, which participate in the same proprietary knowledge economy as the pharmaceutical industry and serve drug companies by harvesting, bundling, and reselling patient data that can be repurposed for drug development or clinical trial recruitment. In the knowledge economy of digital health, the patient's voice becomes knowledge only by being disaggregated as voice and reaggregated as data that can be accumulated at scale and mined by proprietary algorithms. What is being sourced on 'crowdsourcing' sites is not users' insights but the processable data trail they leave in interacting with digital platforms.

RxISK, by contrast, is designed to interpellate the patient as a genuine subject of knowledge about drugs and their effects. The site's main homepage icon is a handheld loudspeaker next to the headline: 'No one knows a prescription drug's side effects like the person taking it.' The patient it addresses is a consumer of drugs, but a creative and critical one, a restless collector of information and a prolific source of novel insights. The same activist outlook is reaffirmed in various forms throughout the site – 'if you think there is a problem, you are probably right', 'Make your voice heard', and so on – in particular on the site's 'About' page or in opinionated blog posts written mostly by Healy himself to provide updates on RxISK's work and commentary on the broader topic of drug-related harms. RxISK's side effect reporting tool is entered through a single click through another prominent home page icon (Figure 5.1). Hence, the entire design and architecture of RxISK.org rests on patient reporting as the cornerstone of side effect research. The contrast is obvious with the FDA's website, for instance, where the reporting system for consumers is grafted on to the agency's reporting system for professionals, and takes clicks through four successive links, several of which are hidden below the fold, to be accessed.[27] (In what is perhaps a subtle sign

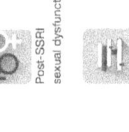

Figure 5.1 RxISK.org home page. The online reporting tool is accessible in a single click through an icon located on the home page (highlighted area in the lower-right corner). So are each one of the site's related functionalities (icons on the middle-left side of the page)

of RxISK's unacknowledged influence, however, a small loud-speaker appeared on the button icon to the FDA's MedWatch reporting form in October 2019, while a clipboard icon, also prominent on RxISK's home page, popped up on the FDA's site in May 2020.[28])

A further difference between RxISK and official pharmacovigi-lance schemes lies in the use that is made of the reports once they are filed. RxISK's reporting tool is an online form that solicits the same sort of information from reporters as the FDA's MedWatch form. In addition to a free narrative account of the reaction, both collect basic metadata needed for the proper storage, retrieval, and interpretation of the report (for example, sex, age, country of residence, name of the suspected medication, dose and duration of treatment, treated condition, concomitant medication, and the like). Yet RxISK's focus on obtaining reports from patients instead of providers reflects a different approach to the analysis of the evidence. Drug agencies typically seek as many reports as pos-sible from manufacturers, prescribers, and consumers alike, and process them in the same manner. Reports filed with the agency are reviewed and encoded, then aggregated into a vast signal-generating database (FAERS for drugs and VAERS for vaccines in the United States).[29] This method explains drug agencies' con-tinued preference for reports by physicians, in which the patient's complaint tends to be filtered and pre-encoded within a stand-ard medical idiom. While the more textured and idiosyncratic reports from patients may be less suited to automated processing, they have other advantages. They focus precisely on those harms which physicians are likely to ignore or dismiss, tend to describe in finer detail what the side effect means to the patient in her everyday life, and in some cases at least have a narrative coher-ence that may facilitate the elucidation of the connection between the effect and its cause. The site's comparatively small scale (it receives approximately two to three thousand reports every year, divided relatively evenly between Europe, North America, and the rest of the world) is consistent with an epistemology that locates the moment of discovery in the close reading of patient self-narratives. Within this framework, the quality of the reports takes precedence over their quantity.

The point of care

In the course of her many forays online, Anne-Marie came across David Healy's name and reached out to him in 2011, shortly before the launch of RxISK.org. He was struck by her story:

> I know a lot about the serotonin system, and here is a woman who at this stage was telling me things about the serotonin system that I didn't know. You had a woman with access to the internet, and motivation to chase things. And she chased a ton of things. Most of what she chased was worthless, but putting one thing into another, taking her time and working slowly, she was able to piece together a story about how SSRIs could be causing the problem, and drugs like mirtazapine might be the answer to an SSRI-induced problem. There's nothing, no articles out there, that actually supported this; there were no doctors who thought this at that point. The only people who agreed with her – they didn't agree publicly – were pharmaceutical companies who were working on drugs like mirtazapine as treatments for alcoholism, but the world didn't know about this.[30]

Persuaded that Anne-Marie's story could serve as a 'symbol [of] what RxISK is all about', he invited her to post an account of her experience on his blog. The post appeared in 2012 and quickly garnered several dozen comments from people who found their experiences reflected in hers. In 2013 Healy collaborated with Anne-Marie to publish a formal report of her case in the *International Journal of Risk and Safety in Medicine*. The following year, two years after RxISK's launch, the same journal published a co-authored paper examining ninety-three further cases of SSRI-induced alcohol dependence drawn from the RxISK database, making these the first two publications to outline this unreported side effect of SSRIs in the medical literature.[31]

One key feature of RxISK illustrated in this story was the use of the internet as a relational technology. Despite the often frustrating nature of their encounters, Anne-Marie found online the resources she needed to keep a conversation going with her doctor. She printed out and brought to him the information that seemed of significance to her, and eventually her quest resulted in the discovery of a suitable treatment. Likewise, RxISK's co-founders envisioned the filing of a side effect report as the beginning of a process that would bring the patient back to their physician. A report filed on RxISK is not merely filed away for later use, as it is in the voluntary reporting systems of national drug agencies, but rather is emailed back to the reporter with a 'RxISK score' that indicates how likely it is that the issue in question is drug-induced (Figure 5.2). The report notes that the RxISK is calculated based on the Bradford Hill causation criteria and is meant to give a preliminary assessment of the connection between drug and reaction. Accordingly, the return email in which the report is included advises the patient to print out the report and bring it to their treating physician:

> We thought that if we could get [patients] to print a RxISK report ... and bring the report to the doctor, it would equalize the power relation a bit. You know if I go in and tell you I'm having a problem and you blow me off and throw me out, then there's no record of it. It's just my word against yours. But if there's a RxISK report brought to the doctor, and they know this has been printed off an expert website and there's a record of it, then the doctor should be less likely to blow you off.[32]

To close the feedback loop, RxISK.org also added in 2017 a portal allowing physicians whose patients brought them a RxISK report to log back into the website and, using a unique identifying code generated for each report, to submit their own observations on the likelihood that their patient's complaint is indeed treatment-induced. The function of the report, therefore, is not merely to move information upwards within a hierarchy of expertise – from patient to doctor to regulator – but also to triangulate the relationship between patient, physician, and website. Implied here is the project to enhance the recognition of side effects not through the fine-tuning of informatic tools for the downstream processing of ADR reports but through a reshaping of the relational complex in which drug-related harms are voiced and recorded in the first place.

This use of the Web to document side effects in a reconfigured set of relations has also informed RxISK's ongoing campaign to highlight the extent of drug-induced sexual dysfunction. SSRIs' ability to disrupt sexual function is familiar to clinicians, though its costs to patients are typically overshadowed. SSRI labels describe changes in sexual function as infrequent, usually mild, and always transitory. In reality, some degree of genital anaesthesia occurs in nearly all patients treated with the drugs. RxISK received several

MAKING MEDICINES SAFER FOR ALL OF US

RxISK Report

Your RxISK score is: **14**. The following is a guide to the score.*

0 - 4 — More information required
5 - 8 — Likely link between medicine and side effect.
9 + — Strong possibility of a link between medication and side effect.

A record of your submission follows.

*Your RxISK score is based on Koch's postulates, the Bradford-Hill criteria, and the Naranjo algorithm for cause and effect. This is a better approach than randomized clinical trials, and is something your doctor or pharmacist should take seriously.

If you provided permission to publish some of your comments from the report, the questions marked [share#] are the ones that may be published. Sharing stories and not just numbers of reports is hugely valuable in allowing others to learn from your experience and understand their own. PLEASE NOTE THAT NO CONTACT DETAILS WILL BE PUBLISHED.

Figure 5.2 RxISK report cover page. A RxISK report is emailed back to anyone who reports a side effect on RxISK. org with the following note: 'This report is designed to be used in conversation with your doctor or pharmacist on the possible linkage between the suspect drug and the primary side effect. You can also invite them to add to your RxISK report to indicate whether they agree that there is a linkage: http://rxisk.org/hcp-comment/.' Accessed 18 May 2018

hundred reports from patients whose drug-induced sexual dysfunction endured after discontinuing treatment with SSRIs, with the hair loss medication finasteride, or with the acne medication isotretinoin.[33] Given its toll on patients' well-being and relationships, enduring sexual dysfunction is a clear example of a drug-related harm of which patients are acutely aware but which they are reluctant to disclose to physicians. To bring the issue to the attention of the FDA, the United Kingdom's MHRA, and the European Medicines Agency (EMA), Healy and Mangin submitted in May 2018 a petition signed by twenty-two specialists of PSSD (post-SSRI sexual dysfunction) and PGAD (persistent genital arousal disorder) requesting a review and revision of the drugs' labels. In conditions of a sensitive or stigmatized nature, low reporting rates and the anonymous nature of most reports tend to compromise the credibility of testimonials. As a way around these obstacles, Healy and Mangin contacted over three hundred individuals who had submitted relevant reports to RxISK to invite them to request supporting documentation from their treating physician and to resubmit their reports with their name and email address to the EMA, indicating in the event their willingness to be contacted. In the short timeframe allotted by the EMA, they were able to submit eighty-two named reports in support of their petition, thirty-two of which contained additional documentation from health professionals. In 2019 the Agency completed its review and ordered labelling changes to reference reports on cases of sexual dysfunction subsisting after discontinuation of the treatment.[34]

Achieving this triangulation between patient, physician, and website is undoubtedly the most elusive of RxISK's goals. As Healy and Applbaum note, the site has had considerably more success in engaging patients than prescribers. In the PSSD/PGAD campaign, documentation from physicians could be produced only by means of a time-consuming outreach effort conducted via patients. Through that campaign, RxISK enacted on an experimental scale what remains at this point a mostly aspirational model for a different way of delivering care. In that regard, RxISK shares with other eHealth sites a distinct promissory dimension; to invite participation from users, it conjures up a vision of a medical future that has yet to come. Its distinctiveness lies in the kind of medical future that is being conjured up, one in which digital tools are used not to bypass the clinical relation but rather to reconfigure it in an effort to move the boundaries of what can be seen or heard when patient and physician meet.

The credibility deficit that affects web content derives in large part from the fact that voice, when disseminated online, is disembodied and disembedded. The speaker herself is no longer present in her voice, but merely represented by it. This moment of representation always involves the possibility of misrepresentation, inauthenticity, or deception. The originality of RxISK is to use the Web not only as a tool for accumulating data but also for reconfiguring relationships across the digital divide. The RxISK report is meant to do more than record information. Its circulation between the parties involved in the therapeutic process fulfils an authenticating function by linking the report to the living and speaking patient who filed it. RxISK's successes in uncovering unrecognized or highlighting downplayed drug harms have involved evidence made usable and credible through this relational approach.

As such, RxISK may be demonstrating the value of online patient reporting for a better understanding of drugs' benefits and harms. But it also demonstrates that, as is the case with the pharmakon, the efficacy of technology is circumstantial; it depends on the manner of its use as much as on its substantive properties. RxISK's reliance on motivated and mobilized patients – those who want to 'make their voices heard' – may explain why SSRIs have continued to loom so large in its work. While RxISK.org invites reports on all prescription drugs, Healy's trajectory put him at the centre of an activist network concerned about antidepressants and their risks and poised to engage with a project like RxISK. This means that RxISK itself may not constitute a panacea, a universal formula to bring the hidden harms of modern pharmacology to light. Nonetheless, in showing that side effects must often be uncovered against or in spite of the normal regime of drug research and regulation, it does demonstrate what role activist communities both on- and offline may have to play in broadening our knowledge of drugs and their effects.

Notes

1 Anne-Marie Kelly, 'Out of My Mind. Driven to Drink.' 15 March 2012. https://ssristories.org/out-of-my-mind-driven-to-drink-davidhe aly-org/. Accessed 30 August 2023.
2 The argument recurs in various forms throughout Derrida's early work. See in particular Jacques Derrida, *Speech and Phenomena*, translated by D.B. Allison (Evanston, IL: Northwestern University Press, 1973), and 'Plato's Pharmacy', in *Dissemination*, translated by B. Johnson (Chicago, IL: University of Chicago Press, 1981), pp. 61–172.
3 For surveys, see Susannah Fox and Lee Rainie, 'Vital Decisions: A Pew Internet Health Report.' 22 May 2002. www.pewresearch.org/internet/ 2002/05/22/vital-decisions-a-pew-internet-health-report/. Accessed 1 October 2022. For early studies, see Michael Hardey, 'Doctor in the House: The Internet as a Source of Lay Health Knowledge and the Challenge to Expertise.' *Sociology of Health and Illness* 21, no. 6 (1999): 1545–53, and ' "The Story of My Illness": Personal Accounts of Illness on the Internet.' *Health* 6, no. 1 (2002): 31–46. A useful overview of the early literature on 'eHealth' is offered in Joëlle Kivits, 'E-Health and Renewed Sociological Approaches to Health and Illness', in *Digital Sociology: Critical Perspectives*, edited by Kate Orton-Johnson and Nick Prior, pp. 213–26 (Basingstoke: Palgrave Macmillan, 2013).
4 The most comprehensive articulation of this argument is Shoshana Zuboff's *The Age of Surveillance Capitalism: The Fight for a Human Future at the New Frontier of Power* (New York: Public Affairs, 2019). Other influential accounts, such as Yochai Benkler, Robert Faris, and Hal Roberts, *Network Propaganda: Misinformation, Manipulation, and Radicalization in American Politics* (Oxford: Oxford University Press, 2018), and Cailin O'Connor and James Owen Weatherall, *The Misinformation Age: How False Beliefs Spread* (New Haven, CT: Yale University Press, 2019), tend to focus more narrowly on the issue of political misinformation, though in the era of COVID-19, the connections between political and medical misinformation have become unmistakable.
5 I have explored aspects of this argument in Antoine Lentacker, 'Epistemology of the Side Effect: Anecdote and Evidence in the Digital Age.' *BioSocieties* 19 (2024): 84–111. I am grateful to Springer Nature for permission to reuse some of the research and language for this article in the present chapter.

6 Greg Andrews, 'Prozac Sales Boom, Near 2 Billion a Year.' *Indianapolis Business Journal* 15, no. 32 (1994): 1A.

7 Food and Drug Administration. 'Psychopharmacological Drug Advisory Committee Hearing.' 20 September 1991, Rockville, MD. https://commons.wikimedia.org/wiki/File:1991_FDA_Psycho pharmacological_Drugs_Advisory_Committee.pdf. Accessed 3 October 2022.

8 John Cornwell, *The Power to Harm: Mind, Medicine, and Murder on Trial* (New York: Vintage, 1996); Karen Barth Menzies, 'A Cure Worse than the Disease.' *Trial* 14, no. 3 (2005): 20–9.

9 David Healy, *Let Them Eat Prozac: The Unhealthy Relationship between the Pharmaceutical Industry and Depression* (New York: New York University Press, 2004).

10 Austin Bradford Hill, 'The Environment and Disease: Association or Causation?' *Proceedings of the Royal Society of Medicine* 58, no. 5 (1965): 295–300.

11 *Forsyth v. Eli Lilly & Co.*, Transcript of Proceedings, case no. 95-CV-00185ACK (D. Haw., 1999), pp. 900–29; *Tobin v. SmithKline Beecham*, Transcript of Trial Proceedings, case no. 00-CV-0025BEA (D. Wyo., 2001), p. 41.

12 Linsey McGoey and Emily Jackson, 'Seroxat and the Suppression of Clinical Trial Data: Regulatory Failure and the Uses of Legal Ambiguity.' *Journal of Medical Ethics* 35, no. 2 (2009): 107–12; David Healy, Joanna Le Noury, and Julie Wood, *Children of the Cure: Missing Data, Lost Lives, and Antidepressants* (Toronto: Samizdat Health, 2020). 'Secrets of Seroxat' can be viewed at https://vimeo.com/ 105150078. Accessed 27 August 2024.

13 See, for example, Andrew Lakoff, 'The Right Patients for the Drug: Managing the Placebo Effect in Antidepressant Trials.' *BioSocieties* 2, no. 1 (2009): 57–71; Steven Epstein, *Inclusion: The Politics of Difference in Clinical Research* (Chicago, IL: University of Chicago Press, 2007); Adriana Petryna, *When Experiments Travel: Clinical Trials and the Global Search for Human Subjects* (Princeton, NJ: Princeton University Press, 2009); Sergio Sismondo, 'Ghosts in the Machine: Publication Planning in the Medical Sciences.' *Social Studies of Science* 39, no. 2 (2009): 171–98; Lochlann S. Jain, 'The Mortality Effect: Counting the Dead in the Cancer Trial.' *Public Culture* 22, no. 1 (2010): 89–117.

14 Jerry Avorn, 'Dangerous Deception – Hiding the Evidence of Adverse Drug Effects.' *New England Journal of Medicine* 355, no. 21 (2006): 2169–71; Harlan Krumholz, Joseph Ross, Amos Presler,

and David Egilman, 'What Have We Learnt from Vioxx?' *BMJ* 334, no. 7585 (2007): 120–3; Kalman Applbaum, 'Shadow Science: Zyprexa, Eli Lilly and the Globalization of Pharmaceutical Damage Control.' *BioSocieties* 5, no. 2 (2010): 236–55.

15 Doron Garfinkel and Derelie Mangin, 'Addressing the Invisible Iatrogenic Epidemic: The Role of Deprescribing in Polypharmacy and Inappropriate Medication Use.' *Therapeutic Advances in Drug Safety* 10 (2019): 1–5.

16 Charles Medawar, 'The Bidet View of Psychopharmacology.' Interview by David Healy, 2002. https://samizdathealth.org/wp-cont ent/uploads/2020/12/Medawar.pdf. Accessed 21 July 2021.

17 Charles Medawar, 'The Antidepressant Web: Marketing Depression and Making Medicines Work.' *International Journal of Risk and Safety in Medicine* 10, no. 2 (1997): 75–126.

18 Charles Medawar, Andrew Herxheimer, Andrew Bell, and Shelley Jofre, 'Paroxetine, *Panorama* and User Reporting of ADRs: Consumer Intelligence Matters in Clinical Practice and Post-Marketing Drug Surveillance.' *International Journal of Risk and Safety in Medicine* 15, nos. 3–4 (2002): 161–9, 167.

19 Medawar et al., 'Paroxetine, Panorama and User Reporting of ADRs', 167–8.

20 Medawar et al., 'Paroxetine, Panorama and User Reporting of ADRs', 161–2.

21 Charles Medawar and Andrew Herxheimer, 'A Comparison of Adverse Drug Reaction Reports from Professionals and Users, Relating to Risk of Dependence and Suicidal Behaviour with Paroxetine.' *International Journal of Risk and Safety in Medicine* 16, no. 1 (2004): 5–19, 15; the emails and their analysis became the subject of a second *Panorama* episode on Seroxat, 'Emails from the Edge', broadcast in May 2003: https://vimeo.com/115681493. Accessed 14 July 2021.

22 Andrew Herxheimer, Rose Crombag, and Teresa Leonardo Alves, 'Direct Patient Reporting of Adverse Drug Events, Briefing Paper.' Amsterdam: Health Action International, 2010. https://consumers. cochrane.org/sites/consumers.cochrane.org/files/uploads/10%20 May%202010%20Report%20Direct%20Patient%20Report ing%20of%20ADRs.pdf. Accessed 21 November 2022.

23 Deborah Lupton, 'The Commodification of Patient Opinion: The Digital Patient Experience Economy in the Age of Big Data.' *Sociology of Health and Illness* 36, no. 6 (2014): 856–69; José van Dijck and Thomas Poell, 'Understanding the Promises and Premises of Online

Health Platforms.' *Big Data & Society* 3, no. 1 (2016): 1–11; Minna Ruckenstein and Natasha Dow Schüll, 'The Datafication of Health.' *Annual Review of Anthropology* 46 (2017): 261–78; Kirsten Ostherr, 'Privacy, Data Mining, and Digital Profiling in Online Patient Narratives.' *Catalyst: Feminism, Theory, Technoscience* 4, no. 1 (2018): 1–24; Niccolò Tempini and Lorenzo Del Savio, 'Digital Orphans: Data Closure and Openness in Patient-Powered Networks.' *BioSocieties* 14 (2019): 205–27.

24 David Healy, Personal Communication, 17 October 2018.

25 Kalman Applbaum, Personal Communication, 23 July 2022.

26 Applbaum, Personal Communication.

27 The link to access the FDA's reporting tool for consumers is, as of this writing: www.accessdata.fda.gov/scripts/medwatch/index.cfm?action= consumer.reporting1. Accessed 3 October 2022.

28 Changes on the FDA's website are recorded through the Internet Archive's Wayback Machine. For the first appearance of the loudspeaker icon, see https://web.archive.org/web/20191024040034/www.fda.gov/ safety/medwatch-fda-safety-information-and-adverse-event-reporting- program. Accessed 28 November 2022; for the clipboard icon: https:// web.archive.org/web/20200502005538/www.accessdata.fda.gov/ scripts/medwatch/index.cfm. Accessed 6 October 2024.

29 See the description in Janet Woodcock, Rachel E. Behrman, and Gerald J. Dal Pan, 'Role of Postmarketing Surveillance in Contemporary Medicine.' *Annual Review of Medicine* 62 (2011): 1–10.

30 Healy, Personal Communication.

31 Onome V. Atigari, Anne-Marie Kelly, Qamar Jabeen, and David Healy, 'New Onset Alcohol Dependence Linked to Treatment with Selective Serotonin Reuptake Inhibitors.' *International Journal of Risk and Safety in Medicine* 25, no. 2 (2013): 105–9; Louise Brookwell, Carys Hogan, David Healy, and Derelie Mangin, 'Ninety- Three Cases of Alcohol Dependence Following SSRI Treatment.' *International Journal of Risk and Safety in Medicine* 26, no. 2 (2014): 99–107.

32 Healy, Personal Communication.

33 David Healy, Joanna Le Noury, and Derelie Mangin, 'Enduring Sexual Dysfunction after Treatment with Antidepressants, 5α-Reductase Inhibitors and Isotretinoin: 300 cases.' *International Journal of Risk and Safety in Medicine* 30 (2019): 167–78.

34 RxISK Blog, 'EMA Acknowledges Persistent Sexual Dysfunction after SSRIs & SNRIs.' 11 June 2019. https://rxisk.org/ema-ackno wledges-persistent-sexual-dysfunction-after-ssris-snris/. Accessed 15 July 2021.

Bibliography

Andrews, Greg. 'Prozac Sales Boom, Near 2 Billion a Year.' *Indianapolis Business Journal* 15, no. 32 (1994): 1A.

Applbaum, Kalman. 'Shadow Science: Zyprexa, Eli Lilly and the Globalization of Pharmaceutical Damage Control.' *BioSocieties* 5, no. 2 (2010): 236–55.

Atigari, Onome V., Anne-Marie Kelly, Qamar Jabeen, and David Healy. 'New Onset Alcohol Dependence Linked to Treatment with Selective Serotonin Reuptake Inhibitors.' *International Journal of Risk and Safety in Medicine* 25, no. 2 (2013): 105–9.

Avorn, Jerry. 'Dangerous Deception – Hiding the Evidence of Adverse Drug Effects.' *New England Journal of Medicine* 355, no. 21 (2006): 2169–71.

Barth Menzies, Karen. 'A Cure Worse than the Disease.' *Trial* 14, no. 3 (2005): 20–9.

Benkler, Yochai, Robert Faris, and Hal Roberts. *Network Propaganda: Misinformation, Manipulation, and Radicalization in American Politics*. Oxford: Oxford University Press, 2018.

Brookwell, Louise, Carys Hogan, David Healy, and Derelie Mangin. 'Ninety-Three Cases of Alcohol Dependence Following SSRI Treatment.' *International Journal of Risk and Safety in Medicine* 26, no. 2 (2014): 99–107.

Cornwell, John. *The Power to Harm: Mind, Medicine, and Murder on Trial*. New York: Vintage, 1996.

Derrida, Jacques. *Dissemination*, translated by B. Johnson. Chicago, IL: University of Chicago Press, 1981.

Derrida, Jacques. *Speech and Phenomena*, translated by D.B. Allison. Evanston, IL: Northwestern University Press, 1973.

Epstein, Steven. *Inclusion: The Politics of Difference in Clinical Research*. Chicago, IL: University of Chicago Press, 2007.

Food and Drug Administration. 'Psychopharmacological Drug Advisory Committee Hearing.' Rockville, MD, 20 September 1991. https://commons.wikimedia.org/wiki/File:1991_FDA_Psychopharmacological_Drugs_Advisory_Committee.pdf. Accessed 3 October 2022.

Forsyth v. Eli Lilly & Co. Transcript of Proceedings, case no. 95-CV-00185ACK, US District Court for the District of Hawaii, 1999.

Fox, Susannah, and Lee Rainie. 'Vital Decisions: A Pew Internet Health Report.' 22 May 2002. www.pewresearch.org/internet/2002/05/22/vital-decisions-a-pew-internet-health-report/. Accessed 1 October 2022.

Garfinkel, Doron, and Derelie Mangin. 'Addressing the Invisible Iatrogenic Epidemic: The Role of Deprescribing in Polypharmacy and Inappropriate Medication Use.' *Therapeutic Advances in Drug Safety* 10 (2019): 1–5.

Hardey, Michael. 'Doctor in the House: The Internet as a Source of Lay Health Knowledge and the Challenge to Expertise.' *Sociology of Health and Illness* 21, no. 6 (1999): 1545–53.

Hardey, Michael. ' "The Story of My Illness": Personal Accounts of Illness on the Internet.' *Health* 6, no. 1 (2002): 31–46.

Healy, David. *Let Them Eat Prozac: The Unhealthy Relationship between the Pharmaceutical Industry and Depression.* New York: New York University Press, 2004.

Healy, David, Joanna Le Noury, and Derelie Mangin. 'Enduring Sexual Dysfunction after Treatment with Antidepressants, 5α-Reductase Inhibitors and Isotretinoin: 300 Cases.' *International Journal of Risk and Safety in Medicine* 30 (2019): 167–78.

Healy, David, Joanna Le Noury, and Julie Wood. *Children of the Cure: Missing Data, Lost Lives, and Antidepressants.* Toronto: Samizdat Health, 2020.

Herxheimer, Andrew, Rose Crombag, and Teresa Leonardo Alves. 'Direct Patient Reporting of Adverse Drug Events, Briefing Paper.' Amsterdam: Health Action International, 2010. https://consumers.cochr ane.org/sites/consumers.cochrane.org/files/uploads/10%20May%202 010%20Report%20Direct%20Patient%20Reporting%20of%20A DRs.pdf. Accessed 21 November 2022.

Hill, Austin Bradford. 'The Environment and Disease: Association or Causation?' *Proceedings of the Royal Society of Medicine* 58, no. 5 (1965): 295–300.

Jain, Lochlann S. 'The Mortality Effect: Counting the Dead in the Cancer Trial.' *Public Culture* 22, no. 1 (2010): 89–117.

Kelly, Anne-Marie. 'Out of My Mind. Driven to Drink.' 15 March 2012. https://davidhealy.org/out-of-my-mind-driven-to-drink/. Accessed 30 August 2023.

Krumholz, Harlan, Joseph Ross, Amos Presler, and David Egilman. 'What Have We Learnt from Vioxx?' *BMJ* 334, no. 7585 (2007): 120–3.

Lakoff, Andrew. 'The Right Patients for the Drug: Managing the Placebo Effect in Antidepressant Trials.' *BioSocieties* 2, no. 1 (2009): 57–71.

Lentacker, Antoine. 'Epistemology of the Side Effect: Anecdote and Evidence in the Digital Age.' *BioSocieties* 19 (2024): 84–111.

Lupton, Deborah. 'The Commodification of Patient Opinion: The Digital Patient Experience Economy in the Age of Big Data.' *Sociology of Health and Illness* 36, no. 6 (2014): 856–69.

McGoey, Linsey, and Emily Jackson. 'Seroxat and the Suppression of Clinical Trial Data: Regulatory Failure and the Uses of Legal Ambiguity.' *Journal of Medical Ethics* 35, no. 2 (2009): 107–12.

Medawar, Charles. 'The Antidepressant Web: Marketing Depression and Making Medicines Work.' *International Journal of Risk and Safety in Medicine* 10, no. 2 (1997): 75–126.

Medawar, Charles. 'The Bidet View of Psychopharmacology.' Interview by David Healy, 2002. https://samizdathealth.org/wp-content/uploads/2020/12/Medawar.pdf. Accessed 21 July 2021.

Medawar, Charles, and Andrew Herxheimer. 'A Comparison of Adverse Drug Reaction Reports from Professionals and Users, Relating to Risk of Dependence and Suicidal Behaviour with Paroxetine.' *International Journal of Risk and Safety in Medicine* 16, no. 1 (2004): 5–19.

Medawar, Charles, Andrew Herxheimer, Andrew Bell, and Shelley Jofre. 'Paroxetine, *Panorama* and User Reporting of ADRs: Consumer Intelligence Matters in Clinical Practice and Post-Marketing Drug Surveillance.' *International Journal of Risk and Safety in Medicine* 15, no. 3–4 (2002): 161–9.

O'Connor, Cailin, and James Owen Weatherall. *The Misinformation Age: How False Beliefs Spread.* New Haven, CT: Yale University Press, 2019.

Orton-Johnson, Kate, and Nick Prior (eds). *Digital Sociology: Critical Perspectives.* Basingstoke: Palgrave Macmillan, 2013.

Ostherr, Kirsten. 'Privacy, Data Mining, and Digital Profiling in Online Patient Narratives.' *Catalyst: Feminism, Theory, Technoscience* 4, no. 1 (2018): 1–24.

Petryna, Adriana. *When Experiments Travel: Clinical Trials and the Global Search for Human Subjects.* Princeton, NJ: Princeton University Press, 2009.

Ruckenstein, Minna, and Natasha Dow Schüll. 'The Datafication of Health.' *Annual Review of Anthropology* 46 (2017): 261–78.

RxISK Blog. 'EMA Acknowledges Persistent Sexual Dysfunction after SSRIs & SNRIs.' 11 June 2019. https://rxisk.org/ema-acknowledges-persistent-sexual-dysfunction-after-ssris-snris/. Accessed 15 July 2021.

'Secrets of Seroxat.' https://vimeo.com/105150078. Accessed 27 August 2024.

Sismondo, Sergio. 'Ghosts in the Machine: Publication Planning in the Medical Sciences.' *Social Studies of Science* 39, no. 2 (2009): 171–98.

Tempini, Niccolò, and Lorenzo Del Savio. 'Digital Orphans: Data Closure and Openness in Patient-Powered Networks.' *BioSocieties* 14 (2019): 205–27.

Tobin v. SmithKline Beecham. Transcript of Trial Proceedings, case no. 00-CV-0025BEA, US District Court for the District of Wyoming, 2001.

Van Dijck, José, and Thomas Poell. 'Understanding the Promises and Premises of Online Health Platforms.' *Big Data & Society* 3, no. 1 (2016): 1–11.

Woodcock, Janet, Rachel E. Behrman, and Gerald J. Dal Pan. 'Role of Postmarketing Surveillance in Contemporary Medicine.' *Annual Review of Medicine* 62 (2011): 1–10.

Zuboff, Shoshana. *The Age of Surveillance Capitalism: The Fight for a Human Future at the New Frontier of Power.* New York: Public Affairs, 2019.

Part III

Co-opting disease, promoting
prevention and healing

6

Sunbeds, dihydroxyacetone (DHA) fake tan, and MelanoTan injections: A history of 'safe' tanning technologies

Fabiola Creed

By the early 1990s, many oncologists, epidemiologists, and dermatologists were alarmed by the sharp rise in malignant melanoma skin cancer rates since World War II in fair-skinned populations across Europe, the United States, Canada, and Australia.[1] In response, the Department of Health in Britain published *The Health of the Nation* (1992), which included a target to stop the year-by-year rise of skin cancers by 2005. The government report was largely concerned with the depleting ozone layer and other cancer-causing 'green issues'.[2] Yet health experts and the media focused on sunbeds because people still used them despite their well-known carcinogenic effects. Also, sunbed use was considered a preventable cause of skin cancer, unlike general sun exposure or broader environmental factors.[3] Therefore, medical authorities – mainly cancer specialists and dermatologists – assumed their discouragement of sunbeds could reduce skin cancer rates.

However, the British government, medical experts, and the media struggled to weaken the sunbed industry's commercial power or remove sunbed technologies from households and commercial spaces.[4] Throughout the twentieth century, most Westerners associated tanned white skin with health, wealth, and athleticism, and paleness with sickliness and poverty.[5] As one dermatologist explained, sunbeds therefore signified 'health by association' within these populations.[6] Several preceding medical technologies had also reinforced the positive associations of a sunbed tan. As scholars Tania Woloshyn and Simon Carter have noted, ultraviolet (UV) lamps were invented as a medical technology in the 1890s to prevent or cure infections and diseases, such as rickets and tuberculosis. Throughout the following century, 'fitness-enhancing' and 'beautifying' variations of UV devices were then sold as a cure for skin conditions, including acne, psoriasis, and eczema, and mental health issues, such as seasonal affective disorder and depression.[7] In 1978, the modern sunbed arrived in Britain, and by the early 1980s, a boom in the interlinked health club and sunbed industry strengthened White people's obsession to develop strong, 'healthy', and tanned physiques.[8] Though sunbeds were not introduced as 'medical' technologies', some general practitioners on occasion prescribed them for skin conditions, and providers claimed that sunbeds were safer than sunbathing outdoors.[9]

After the mid-1980s, however, concerns about sunbeds overshadowed claims of their virtue. In the mainstream media, most presses removed sunbed advertisements from their pages.[10] Nonetheless, sunbed companies continued spreading their own promotional material, using the same misleading yet long-established claims that sunbeds were safe.[11] Medical studies, media coverage, and even government campaigns did not lessen UV consumption either.[12] Most people continued to use sunbeds, albeit discreetly, to avoid shame.[13] In Britain, most White people still desired tanned skin, and although sunbed technologies were no longer novel or fashionable, they proved an effective way for sun-starved people to develop an all-year tan. In the early 1990s, skincare providers and scientific research institutions soon realized they could exploit this consumer demand by providing advanced 'UV-free' tanning technologies.[14] More importantly, they could use the ongoing UV-induced skin cancer concerns to seek both medical and media endorsement. By the early 1990s, all media coverage on UV-free tanning technologies emerged in the spaces where sunbeds had once featured.

This chapter focuses on two of these 'UV-free' tanning technologies: first, the updated formulas of dihydroxyacetone (DHA) tanning – through lotions, creams, mousses, and sprays – and then the entirely novel invention of MelanoTan injections.[15] To expose ongoing issues with DHA serums since their mid-1950s discovery, I examine consumer reviews from the *Financial Times* newspaper, and the *Which?*, *The Druggist and Chemist*, *She*, and *Cosmopolitan* magazines. The DHA serums of the 1990s were apparently improved versions of past products; the new products could be applied at home and, for the first time, by tanning specialists in beauty salons. Scientists also began to develop other tanning methods funded by their research institutions. Reportedly, MelanoTan was an attempt to create a medical technology that both darkened skin and protected people from skin cancer. In this chapter, I will interchange the terms 'fake tan', 'self-tanning', and 'alternative tanning' when discussing UV-free technologies, and do so neutrally. In the twentieth century, the British media and public used the term 'fake' to derogatorily describe a tan that had not developed from UV exposure – even though other journalists were trying to promote UV-free products.

By historicizing these two very different self-tanning technologies, this chapter demonstrates how fake tan industries both responded to and sought to profit from skin cancer anxieties concerning mainly White people in Britain. As print press coverage and medical journals show, such industries and the media introduced these technologies as an entirely 'safe' alternative to sunbeds. However, I argue that the UV-free tanning industries counterintuitively promoted sunbathing and sunbed use and revived overall tanning culture;[16] they tried to poach the sunbed industry's original client base and attract new consumers with no previous tanning interest. Moreover, the visual and rhetorical strategies employed by the industries revealed the extent to which stakeholders both praised the 'natural'- and ridiculed the 'fake'-looking tan. As beauty editors and consumer critiques explained in the print press, DHA serums, unlike sunbeds, were still in the early stages of product development and thus did not have a favourable reputation for providing a fail-safe 'natural'-looking tan.[17] Finally, the threat posed by alternative tanning products also inspired sunbed franchises to improve their technology, with all industries downplaying the dangers and amplifying health claims to sell their products. As such, the attempts to reduce skin cancer rates through the invention of 'safe' technologies instead likely revived UV tanning in both practice and appeal.

Some historians may argue that a study on tanning products does not belong in the history of technology or medicine; however, a 'technology' is the reciprocated application of knowledge to develop an 'innovative invention' – or, as historian David Edgerton prefers, simply a 'new thing' – which, in this chapter, consists of two consumer products that can change a person's skin colour without UV exposure and therefore affect many people's lives and health.[18] As this chapter illustrates, the health and beauty industry, medical experts, the media, and everyday people worldwide have deemed tanning technologies, like others, 'necessary', 'revolutionary', and 'successful' at certain points in time.[19] Therefore, their history must be critically assessed. Equally, these technologies speak to the varied categorization of the patient consumer.[20] Although the chapter in some senses focuses on beauty consumers, some people were 'health' consumers and patients who self-prescribed sunbed 'treatments' for self-diagnosed skin conditions and mental health issues. Others consumed tanning technologies operated or administered by 'healthcare' providers within 'well-being' spaces. Moreover, alternative tanning technologies were either invented or marketed as an option for patients with pigment-based skin issues who were more susceptible to skin cancer. And, finally, some sunbed and MelanoTan consumers eventually became patients because of the harmful effects of these technologies.

In addition to broadening our view of patient consumers and medical technologies, a focus on new tanning technologies of the 1990s will contribute to historiographies on harmful health and beauty technologies and how such technologies – often invented to reduce 'risk' – can instead create new kinds of health issues.[21] Scholars such as Virginia Berridge and Penny Starns, for example, show how innovations, such as the low-tar cigarettes of the 1970s and the medicalized e-cigarettes of the new millennium, were initially introduced as 'safer cigarettes'. Yet these 'healthier alternatives' were eventually proved to be comparably carcinogenic, or product failure led to insufficient consumer demand. Subsequently, consumers went back to their former ways of smoking, while younger generations had new nicotine addictions to satisfy.[22] Like the history of UV and UV-free tanning industries, none of these 'advanced' technologies truly encouraged cessation, which would have been the safest but a profitless solution. As such, this chapter also demonstrates how commercial industries have increasingly taken the lead to 'resolve' – and often capitalize from – public health concerns.[23] Moreover, as the *Daily Mail* sources will show, the sunbed industry deployed similar advertising tactics to the cigarette industry.[24] Indeed, the mass media was more than a means for circulating the interests of industry, medical, and other non-media groups; journalists also actively promoted and were excited about new technologies. They wanted to both trial new products for themselves and use their reviews of 'revolutionary' products to attract new readers.

Finally, the history of tanning, alternative tanning products, and skin cancer in this period echoes other cancer and public health histories, supporting that developed countries primarily focused on cancer prevention technologies for White people.[25] Fake tan products and sunbeds also catered to the same market in a mainstream beauty culture that largely presumed Whiteness. This chapter will focus on this beauty culture and target audience in the second half of the twentieth century, with a particular focus on the 1990s. Although it is beyond the chapter's scope, it is worth noting that young White women and men were not the only avid tanning consumers after the twentieth century. People of all genders, sexualities, socioeconomic backgrounds, and now most ethnicities and races have become tanning consumers in Britain. Nowadays, with growing tanning companies, cultures, and influencers, increasing numbers of Brown and Black people are told that darkening their skin will make them 'flawless', 'healthier', look more athletic and slimmer, and, therefore, 'happier' – even if it could one day lead to cancer.[26]

The invention and evolution of DHA 'sun-less' serums

In the mid-1950s, the American scientist Eva Wittgenstein accidentally discovered that the chemical DHA – a white crystalline powder – could artificially tan skin. Working at the Children's Hospital at the University of Cincinnati, Ohio, Wittgenstein was trying to find a treatment for children suffering from a rare metabolic disorder. The children's bodies could not store glycogen, a vital carbohydrate for the body, and DHA was a carbohydrate produced in animals and plants such as sugar cane. In an experiment, Wittgenstein orally administered DHA to children, which caused some to vomit. The experiment had failed, but a few hours later, their skin turned brown where the DHA had fallen. The DHA (a carbohydrate) reacted with the top layer of skin (mostly protein), causing the 'Maillard' reaction. Wittgenstein subsequently tested an aqueous solution of DHA on her own skin, which also turned brown.[27]

In Britain, growing numbers of people sought to tan their skin, a development that was later boosted by the discovery of DHA. Cheap holiday packages to warmer climates, which included air travel, emerged in 1955. From the 1960s onwards, the holiday industry grew over time as travel to Spain and Italy became cheaper than staying in Britain.[28] The subsequent growth in tanning consumerism was particularly noticeable in renowned women's and men's fashion magazines, such as Britain's *Vogue* and the *Gentlemen's Quarterly* (*GQ*). In the 1950s the front covers of *GQ* already featured rugged, middle-aged, extremely tanned, and adventurous-looking White men. By comparison, White women on the front covers of *GQ* and *Vogue* only became bronzed from 1961 onwards – previously, they were all 'ivory' white.[29]

Only the previous year, however, in the summer of 1960, new DHA 'Night Tan' serums had reached people in Britain through extensive national advertising. In the spring of that year, Ellanby Laboratories, a subsidiary of Lewis and Burrows based in London, was apparently the first to launch their new 'Night Tan' clear solution for women.[30] The United Kingdom-originating 'He-Tan' for men appeared in trade magazines that same spring. By the summer, Rolls Razor, also in London, released the still famous 'Man-Tan'. Both 'He-Tan' and 'Man-Tan' were advertised as an aftershave lotion and an indoor tanning cream.[31] The Consumers' Association wanted to publish consumer tests on 'Night Tan' and 'He-Tan' by the end of summer; however, the flood of new DHA serums that season meant they had to start again. The Consumer Association's magazine, *Which?*, published the final report in December 1960. The magazine also refocused on the 'Tanfastic' brand because their cream applied a more even colour when compared to the clear liquid solutions on the market.[32]

As explained in the report, the Consumer Association found that users and the cosmetic industry were still 'not completely satisfied' with even Tanfastic's results.[33] In particular, the 'stain' of a tan was supposed to develop within twenty-four hours; yet the cream did not produce both an all-over and all-even 'natural' tan on half of the participants. One-third perceived the shade as either 'yellowish' or 'orange or red'. Roughly half found the results streaky and patchy, and some had allergic reactions to the DHA or other ingredients, including the added perfume. Others found that the results faded quickly, and the serums stained their clothes, bed-clothes, hair, and beards.[34] Within a few weeks, the more wide-spread *Druggist and Chemist*, a British weekly print magazine for community pharmacists and pharmacy staff, published *Which?*'s disappointing findings.[35]

Nonetheless, the combined print press advertising expenditure of £140,000 across all DHA providers that year – mainly in the summer – still led to the industry's first, yet short-lived, 'DHA boom'. When first released during Britain's 'sunless' summer, people were compelled to seek fake tan preparations.[36] Yet the combination of poor user experience, negative press, and then the winter season led to a loss in consumer interest and sales at the end of the year. Subsequently, in 1961, sales were only 50 per cent of the previous year as DHA tanning was too expensive for its now well-known 'unsatisfactory' results. In Britain, most DHA tanning industries soon disappeared from lack of demand, but Ellanby's 'Night-Tan' and Ambre Solaire's 'Golden', 'Man-Tan', and 'Tanfastic' just about survived, and the DHA industry stabilized at £400,000 per year.[37] This short-lived DHA boom in 1960, and the surviving companies throughout the 1960s, eventually led to DHA becoming a much cheaper 'basic' cosmetic ingredient for a future of self-tanning.

At the beginning of the 1970s, companies claimed they had resolved their 'teething problems' through new and improved DHA products, which further propelled the industry.[38] For example, 'Tanfastic' introduced 'Tanfastic Extra' to their DHA range, which 'kept you browner, [for] longer'.[39] Technological developments also meant that consumers could spread the mixtures on to their skin more easily through a broader range of mousse aerosols, tubed creams, bottled lotions, and 'unbreakable' polythene 'squeeze' packs. The novelty and promise of these innovations increased the value of the DHA tanning market to roughly £600,000 by 1971.[40] While consumer testing had long been underway, the American Food & Drugs Administration (FDA) only added DHA to their list of permanently approved cosmetic ingredients, deeming the external application as safe in 1973.[41] Still, by the mid-1970s, consumers were still complaining about the 'fallibility' and subsequent 'tiger stripes' of even these newly improved formulas.[42] Nonetheless, medical experts still endorsed DHA tanning and even advised vitiligo patients to use DHA products to help cope with skin depigmentation.[43]

During the 1970s and 1980s, the producers of new 'Duo' serums also instructed users to apply self-tans before, during, and after holidays as a protector against UV rays.[44] Erroneously, these advertisements taught people that DHA serums offered strong UV protection. Consequently, when panics about sunbeds and skin cancer arrived, companies confidently asserted deep-rooted claims of UV protection to advance the second 'boom' of DHA tanning.

The DHA tanning technologies of the 1990s

In 1991 the *Daily Mail* was one of the first newspapers to nationally introduce in-depth articles on 'new' and 'improved' DHA serums in Britain. Throughout the decade, the newspaper – especially on its 'Femail' pages – published the largest volume of both tanning articles and advertisements compared to any other British newspaper. In 1992 alone, when these tanning articles featured, roughly 1.7 million *Daily Mail* papers circulated per year in Britain.[45] As the largest circulated tabloid other than the *Sun*, it targeted working- and middle-class men and women.[46] Although it had one of the largest female readerships of any other newspaper,[47] its 'Femail' section both catered for and was regularly read by men.[48] The *Daily Mail* was the most up to date with tanning technologies and was one of the most accessible and widely read sources of tanning information for consumers.

The *Daily Mail* articles on tanning technologies were also very comprehensive. Typically, reporters conducted interviews with tanning specialists and scientists and shared their 'expert' knowledge with the public. As 'consumers' themselves, style and beauty editors also provided their own tanning product critiques. Collectively, these medical and consumer experts wanted to evoke sunbed anxieties – and reduce consumption – through their personal support of DHA serums and other future tanning technologies.

The first *Daily Mail* article on DHA, for example, began with their beauty editor's personal anecdote of her first DHA self-tan experience back in 1981, when she was fifteen years old; Newby Hand's skin had turned a 'patchy, pale yellow' three hours after applying the cream. The following morning, she was 'orange'. Hers was such an unwelcome experience that she avoided fake tan for the following decade. In 1991, however, she became aware of the 'hazardous effect of even short-term sunbathing', which encouraged her to finally trial a few 'dramatically' improved and 'totally safe' fake tans; these included 'Clarins Self Tanning Sun Wrinkle Control Cream (75ml)' for her face and 'Clarins Self Tanning Milk (125ml)' for her body.[49]

Clarins, a French luxury company, was at the forefront of several renowned companies that had newly developed their DHA serums to help discourage UV tanning. In the United States, the president of the Skin Cancer Foundation (a dermatologist, Penny Robins) and the vice-president of research and development for sun pharmaceuticals (Chris Vaughan) strongly endorsed the improved formulas of DHA self-tanners as the only suitable way to tan, and Clarins was, again, the first self-tan to be recommended.[50] Clarins sold DHA serums nationwide in Britain – along with their other skincare, cosmetic, and perfume products – through high-end department stores and salon counters. Clarins had launched this self-tanning face cream and updated 'milk' preparation (both for £9.75) the previous year, in January 1990. This original retail price was high for such a small amount and increased quite quickly for the first few years, suggesting commercial success.[51] Before Hands trialled these products, she interviewed a skin beauty specialist employed by Clarins. The specialist explained how the serum functioned, emphasizing its 'harmless[ness]' because it only stained the top two to three 'superficial' layers of skin for a few days.[52]

Next, the reporter interviewed a technical development advisor employed by L'Oréal. L'Oréal, another French company, is still one of the largest international cosmetics, skincare, sun protection, make-up, and perfume providers. L'Oréal's advisor explained why their new tanning serums were superior to their 1980s predecessors. The developers had now 'stabilised' the DHA and added 'good quality moisturisers'. This apparently stopped skin from turning 'yellow' and made their self-tan 'easier to apply'. Both companies had also masked DHA's ill-reputed 'metallic' smell with different perfumes.[53]

Hands and the tanning 'experts' explained how DHA serums were a safe alternative to sunbeds. Yet the final paragraph of Hands' article weakened the power of this message. Hands advised 'natural' tanning before applying a serum to achieve the top result. She concluded that DHA products 'work[ed] best when used over an existing (real) tan, no matter how light'.[54] She did, however, instruct readers to use high sun protection and limit their sun exposure hours. Moreover, the image chosen for this article featured a White woman on a sandy beach, dressed in a black and white swimsuit to emphasize her tan. The model smiled with her face tilted upwards, lit by the sun.[55] In the 1990s these types of images were featured in every newspaper article warning against UV tanning. Clearly, anti-UV rhetoric did not preclude the constant visual glorification of the sun or the happiness that came from achieving a 'healthy' tan.[56]

In 1992 the British public apparently spent £3.6 million on 'self-tanning' products alone. The following year, self-tanning became the 'fastest growing sector in the sun care market'.[57] To capitalize further, the application of DHA serums shifted in part from domestic spaces to beauty salons, and appliers were now skilled beauty experts who had undergone rigorous training.[58] One of these experts, a Decléor beauty therapist and beauty salon owner, was interviewed in 1993, again for the *Daily Mail*. Her clients used her DHA tanning services because they wanted the 'psychological and aesthetic benefits of a light tan' without the damage from UV exposure.[59] In the same article, the reporter interviewed a spokeswoman for Ambre Solaire, the sun care strand of Garnier. Both Decléor and Garnier were skincare brands owned by the French cosmetics company L'Oréal. The spokesperson boasted that Ambre Solaire now provided three more shades to match better the 'natural tan' of different skin types. The reporter, Katie Hayward, underwent this 'Decleor self-tanning treatment' herself to explain the long process to her readers. She had to undergo a thorough, twenty-five-minute full-body exfoliation to avoid 'blotchy' results. After a shower, the first tanning application took thirty minutes. After five hours, Hayward had to return to the salon for another full-body application. Hayward's fake tan lasted for over a week, but it cost a total of £63, whereas one sunbed session cost a maximum of £4. As such, this alternative tanning was clearly not accessible to all or even most people in Britain.[60]

All DHA tanning coverage also slandered pale white complexions in the 1990s. Even though the media warned people about the ageing and skin cancer effects of sunbeds and sunbathing, people were instructed on how to avoid 'pallid' skin before and after sunbathing holidays.[61] *She*, *Company*, *Vogue*, and *Marie Claire* strongly endorsed DHA serums as a safe alternative to sunbeds.[62] But soon, most reviews by both reporters and everyday consumers explained how these 'improved' mixtures were still not quite right. Their skin turned yellow, orange, blotchy, patchy, streaky, uneven, or the shade was 'too light' or 'too dark'; the textures were sticky, runny, or greasy, and, finally, the results washed off easily or were too expensive.[63] Some people even criticized the relentless pursuit of trying to find a 'natural'-looking fake tan more generally,[64] and focus groups still reported the stench of serums.[65] As self-tan producers and journalists reported, this alone drove people back to sunbed tanning even though they had fallen out of fashion by the late 1980s.[66] One man even asserted that sunbeds remained the most 'acceptable' and 'popular' tanning method for men, whereas the external application of a cosmetic product was often deemed 'unacceptable' as it was feminized.[67] Moreover, following *Baywatch* (1989 to 1999), even smaller swimwear and sportswear became fashionable, and the additional skin exposure encouraged full-body 'natural' tans. Viewers were mesmerized by Pamela Anderson's permanently bronzed skin on *Baywatch*, but without the Californian sun to develop it, they sought other failsafe – sunbed – means in colder Britain.[68]

Forecasting the future of 'medical' tanning injections

In 1993 another anti-sunbed, yet ironically pro-tanning, reporter from the *Daily Mail*, Louise Atkinson, published some of the first detailed articles on tanning injections. The government and medical organizations demanded this research following concerns with the depleting ozone layer, the improvements in UV damage detection technology, the rising prevalence of skin cancer in some countries like Australia – where melanoma had overtaken bowel cancer as the most common cancer – and, finally, the observed resilience of tanning culture among White people in Britain. Yet again, the media, the commercial industries, and even the scientific researchers themselves counterintuitively idolized bronzed white skin in all their MelanoTan discussions; they believed people would continue to desire tanned skin in the future. As such, in laboratories around the world 'scientists [began] working on more complicated and infinitely more effective routes to [achieve] that elusive perfect fake tan'. Atkinson asserted that these 'safe and realistic instant tan' injections would soon become a 'dream come true' for the 'typical pasty-white English rose'.[69]

In the first half of her newspaper article, Atkinson explained how scientists at the University of Arizona were conducting 'radical research' by testing a 'melanocyte stimulating hormone' on animals. Scientists had created a synthetic version of this hormone, called MelanoTan, which they tested on pale-skinned men. After ten days of daily injections, the men developed a tan on their heads and shoulders. According to the scientists, the purpose of the research was to create an artificially induced tan to 'protect vulnerable, fair skin[ned]' people against the risk of skin cancer. The inventor, endocrinologist Professor Mac Hadley, explained how MelanoTan functioned by 'closely mimicking the body's natural tanning process, tricking the pigment cells into behaving as they do in the sun'.[70] Professor Mac Hadley started this research on melanocytes (a melanin-forming cell, mainly in the skin), alpha-melanocyte-stimulating hormones (A-MSH) (a hormone that reduces food intake and energy expenditure), pigmentation, and melanoma cancer in the 1980s.[71] When Atkinson was writing this article, these scientists were adhering to the FDA guidelines by slowly increasing MelanoTan's concentration to darken their test subjects more evenly. If MelanoTan were successful, Hadley stated that it would medically treat pigment-based skin problems, such as 'hypersensitivity, albinoism, and vitiligo'.[72] Hadley predicted that within three years, MelanoTan would be on the market, at first through prescriptions to people in Britain who were most susceptible to sunburn and skin cancer. Atkinson's interview with Hadley suggested that the purpose of MelanoTan was explicitly 'preventative' and 'medical' in nature.

Yet, not everyone was so optimistic about the prospect of tanning injections. The second half of Atkinson's article shared the opinion of a MelanoTan sceptic, Professor Patrick Riley, an expert on pigmentation at University College London.[73] Riley started his research career in the 1960s, researching the effect of different environmental factors on the formation of cancer. Like Hadley, he spent much of his career researching melanocytes and melanoma.[74] Sceptical of MelanoTan, Riley argued that too many other factors affected pigmentation, such as skin thickening, keratin levels, and the nervous system, which collectively influenced the skin's protection properties. He argued that the value of MelanoTan was 'purely cosmetic', not medical, illustrating the constant difficulties in separating the medical, commercial, and aesthetic 'purposes' of tanning technologies. Nonetheless, Atkinson concluded that tanning injections would reach the British market and help overcome the 'undisputed hazards' of sun exposure and sunbeds.[75] Yet again, positioned between the headings 'Good Health' and 'Tomorrow's Tan', another accompanying photograph counterintuitively captured a swimwear model enjoying the sunrays.[76]

In newspaper articles, even if the objectives and science behind the new technologies were misinterpreted or disingenuous, print press coverage both reflected and reinforced the importance of tanning culture for medical authorities, scientists, the media, and everyday people in Britain. These stakeholders would rather fund, create, publicize, and demand technologies that enabled 'safer' tanning than abandon the relentless pursuit of maintaining a bronzed complexion. Several reporters, like Atkinson, promoted that these future technologies would offer skin cancer prevention and, in turn, alleviate the public health pressures weighing on people in Britain. In other words, the articles asserted that people would not have to give up suntanning in the future.

During the late 1990s, Professor Hadley began MelanoTan injection trials, in part funded by the Australian government. The side effects from the clinical trials demonstrated the impracticability of MelanoTan. In the trials, people experienced loss of appetite, drowsiness, nausea, vomiting, and erectile complications for men. Scientists were still researching MelanoTan in 2004, aiming to launch the product by 2006.[77] However, by the mid-2000s, MelanoTan injections were illegally sold through the internet, as well as in some tanning salons and bodybuilding gyms in Europe, Australia, and the United States.[78] The clinical trials revealed that MelanoTan should not be used, even when medically monitored by experts. From 2007 onwards, international health agencies and the Medicine and Healthcare Products Regulatory Agency (MHRA) in Britain started warning against its use. The added risk of needle cross-contamination was another major concern.[79] To this day, MelanoTan remains a growing health issue. Consumers now ironically combine MelanoTan with the very technology it was supposed to replace – that is, sunbeds – in the pursuit of an even darker 'natural' tan.[80]

Conclusion

Although fake tanning denigrated the sunbed industry somewhat, and concerns with sunbed-induced skin cancer rose, these alternative tanning industries, the media, and scientists collectively revived the appeal of sunbeds and overall UV exposure by the end of the twentieth century. In all print press coverage and on everyday television, visuals of glamorous models suntanning on beaches undermined any accompanying UV warnings. Moreover, the print press told people that DHA serums, alongside other 1990s tanning technologies,[81] functioned better before or after UV exposure. As earlier versions of DHA serums included a 'satisfactory' Sun Protection Factor (SPF), people were also under the false impression that their skin was protected from UV radiation once artificially bronzed. Throughout all media coverage, people were also shamed for having pale white skin, and told that 'natural' tans were superior to all 'fake' tans. All non-UV tans were described as 'fake' – even if people could not differentiate. In reaction, people would develop entirely new sunbed habits or gravitate back to sunbeds if they had an unfortunate 'fake' tan experience.[82]

At the same time, the sunbed industry was developing a more resilient, low-risk, and quick-spreading franchising approach to thwart this new alternative tanning competition. By 1994, The Tanning Shop became one of the largest national chains in Britain, placing sunbed salons in league with the ubiquitous McDonalds fast food restaurants; sunbeds were cheap, abundant, and accessible for everyone in Britain.[83] The Tanning Shop had also developed technologically improved and purportedly 'safer' vertical 'Hex Honeytan' sunbeds to replace their old horizontal machines. The advertisements for these new models were strategically placed underneath those for DHA tanning salons.

The Tanning Shop also promoted their stress-free sunbed sessions; a session now only took six minutes (and cost £4) instead of thirty–sixty minutes. This contrasted with the expensive Decléor self-tanning treatment mentioned above, which took seven hours from start to finish.[84] Nonetheless, some people combined both UV-free and UV tanning to develop quicker, 'deeper', and longer-lasting tans.[85] Moreover, sunbed providers continued to advertise the 'health benefits' of using their products, including them being a source of vitamin D, and treatment for both skin conditions and depression. Like the selling of cigarettes to women in twentieth-century Britain, most tanning advertisements sold a lifestyle of happiness, youth, vitality, and enhanced sexuality to White viewers if they developed their own 'natural' tans.[86] By 1999, the *Health Education Authority* and *The Times* worryingly estimated that approximately three million people in Britain continued using sunbeds every year.[87] Later, in the mid-2000s, the illegal sale of MelanoTan online and in gyms and tanning salons encouraged the consumption of these products in combination with sunbeds, and likely initiated another revival of tanning culture. The Tanning Shop, for instance, is still a successful franchise to this day.

Collectively, both DHA and MelanoTan technologies perpetuated tanning culture and inadvertently prompted the suntanning public to consider combining both products with sunbed use and sunbathing. One of the apparent objectives – to invent technologies to reduce UV-induced skin cancer – had clearly failed, and instead it helped grow people's obsession with tanning in Britain. Meanwhile, the other objective – to capitalize on skin 'darkening' desires – expanded to people of all genders, sexualities, ages, classes, and most ethnicities and races, as a wider array of people started using these tanning technologies by the early twenty-first century in Britain. Although fake tan products and advertisements initially catered to a mainstream beauty culture that prioritized Whiteness, tanning companies and culture ultimately aimed to boost sales and gain followers. The number of tanning technologies and patient consumers continues to grow into the twenty-first century, further entangled in the complicated histories of ethnicity and particularly Black beauty politics in Britain.[88] As Nina Jablonski – an anthropologist of human skin colour – predicted just ten years ago, 'methods of changing skin color, whether permanently or semi-permanently, will become [even] more varied and sophisticated'.[89] Though the future is unclear, the advertising of technologies that promise to achieve these goals will no doubt include medicalized health claims. Like past tanning technologies, they will also pose new health risks to younger generations and, therefore, an ever-expanding range of patient consumers.

Notes

1 G. Severi, G.G. Giles, C. Robertson, P. Boyle, and P. Autier, 'Mortality from Cutaneous Melanoma: Evidence for Contrasting Trends between Populations.' *British Journal of Cancer* 82, no. 11 (2000): 1887–91.
2 Department of Health. *The Health of the Nation – A Strategy for Health in England* (London: Her Majesty's Stationery Office, 1992); David J. Hunter, Naomi Fulop, and Morton Warner, *From 'Health of the Nation' to 'Our Healthier Nation'* (Copenhagen: World Health Organization, Regional Office for Europe, Policy Learning Curve Series, no. 2, August 2000), p. 3.
3 David Shuttleworth, 'Sunbeds and the Pursuit of the Year Round Tan Should Be Discouraged.' *British Medical Journal* 207 (1993): 1508–9.
4 Fabiola Creed, 'From "Immoral" Users to "Sunbed Addicts": The Media-Medical Pathologising of Working-Class Consumers and Young Women in Late Twentieth-Century England.' *Social History of Medicine* 35, no. 3 (2022): 770–92.
5 For a history of why Western White people sought to develop tanned skin, see Nina Jablonski, *Living Color: The Biological and Social Meaning of Skin Color* (Berkeley, CA, Los Angeles, CA, and London: University of California Press, 2012), pp. 168–79; Devon Hansen, 'Shades of Change: Suntanning and the Twentieth-Century American Dream' (PhD dissertation, Boston University, 2007); Sally Romano, 'The Dark Side of the Sun: Skin Cancer, Sunscreen, and Risk in Twentieth-century America' (PhD dissertation, Yale University, 2006); Kerry Segrave, *Suntanning in 20th Century America* (Jefferson, NC: McFarland & Company, Inc., 2005); Daniel Freund, *American Sunshine: Diseases of Darkness and the Quest for Natural Light* (Chicago, IL: University of Chicago Press, 2012).
6 Shuttleworth, 'Sunbeds and the Pursuit of the Year Round Tan Should Be Discouraged', 1508–9.
7 Tania Woloshyn, 'Le Pays du Soleil: The Art of Heliotherapy on the Côte d'Azur.' *Social History of Medicine* 26, no. 1 (2013): 74–93; Tania

Woloshyn, *Soaking Up the Rays: Light Therapy and Visual Culture in Britain, c. 1890–1940* (Manchester: Manchester University Press, 2017); Simon Carter, *Rise and Shine: Sunlight, Technology and Health* (Oxford: Berg, 2007).

8 Fabiola Creed, *The Rise and Fall of the Sunbed in Britain: Tanning Culture from Fad to Fear* (London: Bloomsbury, 2025).

9 Anon., 'COMPANY COUNSEL.' *Company*, October 1985, p. 190; Lucia van der Post and Joan Price, 'Browned Off without Sun.' *Financial Times*, 28 June 1980, p. 13.

10 I quantified the numbers of sunbed advertisements within the *Daily Mail, The Guardian, The Observer, The Times, The Independent*, and the *Financial Times* from the 1970s to the 1990s.

11 Anon., *Domestic Appliances and Personal Care Products*. Philips English Catalogue. Eindhoven: Philips International Company Archive, February 1997.

12 I.H. Cameron and Christine McGuire, 'Are You Dying to Get a Suntan?' – Pre- and Post-Campaign Survey Results.' *Health Education Journal* 49, no. 4 (1990): 166–70.

13 Jenny Hope, 'Fear Mounts as Sunbed Fans Ignore the Health Risks.' *Daily Mail*, 5 October 1990, p. 23.

14 UV-free tanning existed long before the 1990s. In the 1930s, liquid stockings and cosmetics for darkening white skin were first introduced in twentieth-century Britain. From the 1950s to the 1970s, women stained their legs with tea bags and gravy browning. White people also consumed Beta-carotene supplements, or a large numbers of carrots, to develop a yellow tone to their skin, Anon., 'Headnotes!', British Pathé Archive, 1939. www.britishpathe.com/video/headnotes/query/fake+tan. Accessed 9 December 2021; Davis, H.J., and P.P.S. (eds), *The Bulletin* 270, no. 270 (Nottingham: Boots Archive, 21 March 1952).

15 During the early 1990s, skin cancer concerns prompted other 'health' industries to both develop and advertise their own alternative tanning products. These included 'tanning boosters', a cream that increased the production of melanin through UV exposure, and, finally, an array of tanning tablets, capsules, and pills, which contained either Canthaxanthin, Beta-Carotene, L-Tyrosine, or Psoralen. Anon., 'TAN NATURALLY WITHOUT THE SUN.' *Daily Mail*, 20 April 1991, p. 15; Anon., 'No Need to Ever Get Sunburnt. Golden Tan. Completely Safe and without the Sun.' *Daily Mail*, 4 June 1992, p. 34.

16 Beatrice Aidin, 'When Faking It Doesn't Cause a Stink.' *Financial Times*, 1 August 2009, p. 37.

17 The 'public', the 'media', 'medical experts', and 'consumers' are not always separate groups of people; people can fit within all these overlapping and interweaving categories. For a more comprehensive understanding of the 'public' in Britain, see Alex Mold, Peder Clark, Gareth Millward, and Daisy Payling, *Placing the Public in Public Health in Post-War Britain, 1948–2012* (London: Palgrave Macmillan, 2019).

18 David Edgerton, *The Shock of the Old: Technology and Global History Since 1900* (London: Profile Books, 2011), pp. ix–xvii.

19 Edgerton, *The Shock of the Old.*

20 Nancy Tomes, *Remaking the American Patient: How Madison Avenue and Modern Medicine Turned Patients into Consumers* (Chapel Hill, NC: University North Carolina Press, 2016); Alex Mold, *Making the Patient-Consumer: Patient Organisations and Health Consumerism in Britain* (Manchester: Manchester University Press, 2015).

21 For a parallel and transnational history of skin lighteners, see Lynn N. Thomas, *Beneath the Surface: A Transnational History of Skin Lighteners* (Durham, NC: Duke University Press, 2020); Ronald E. Hall (ed.), *The Melanin Millennium: Skin Color as 21st Century International Discourse* (Dordrecht, Heidelberg, New York, and London: Springer, 2013); for histories of other harmful beauty technologies, see Sheila Jeffreys, *Beauty and Misogyny: Harmful Cultural Practices in the West* (London: Routledge, 2005); Kathy Peiss, *Hope in a Jar: The Making of America's Beauty Culture* (New York: Metropolitan Books, 1998); Elizabeth Haiken, *Venus Envy: A History of Cosmetic Surgery* (Baltimore, MD: Johns Hopkins University, 1997); Thomas Schlich and Ulrich Tröhler (eds), *The Risks of Medical Innovation: Risk Perception and Assessment in Historical Context* (London and New York: Routledge, 2004).

22 Virginia Berridge, 'Electronic Cigarettes and History.' *Lancet* 383, no. 9936 (2014): 2204–5; Virginia Berridge and Penny Starns, 'The "Invisible Industrialist' and Public Health: The Rise and Fall of "Safer Smoking" in the 1970s', in *Medicine, the Market and Mass Media: Producing Health in the Twentieth Century*, edited by Virginia Berridge and Kelly Loughlin, pp. 172–91 (Abingdon and New York: Routledge, 2012).

23 Berridge and Loughlin, *Medicine, the Market and Mass Media.*

24 Katie Hayward, 'A Golden Tan – and No Streaks' and Amanda Sills, 'The Six-Minute Tanning Chamber', *Daily Mail*, 6 May 1993, p. 36; Berridge and Loughlin, *Medicine, the Market and Mass Media*, pp. 172–92.

25 In the 1990s the Office for National Statistics (ONS) Omnibus Surveys for Britain did not question 'naturally black or brown skinned' people about sunburning or sunbed use. This suggests that government researchers did not want to research UV exposure beyond White ethnic groups. They perhaps assumed that racialized groups could not develop skin cancer or they wanted to avoid racialized medical and political tensions in their development of new public health considerations or approaches. For a twentieth-century history of cancer as a 'white woman's nemesis' in America, see Keith Wailoo, *How Cancer Crossed the Color Line* (Oxford: Oxford University Press, 2010); A. Bulman, 'Letters. People Are Overusing Sunbeds.' *British Medical Journal* 310, no. 6990 (1995): 1327; Office of Population Censuses and Surveys, Social Survey Division. *OPCS Omnibus Survey, UK Data Service, SN: 3739, January 1995.* 1998. http://doi.org/10.5255/ UKDA-SN-3739–1. Accessed 25 January 2020.

26 Paul Deslandes, *The Culture of Male Beauty in Britain: From the First Photographs to David Beckham* (Chicago, IL, and London: University of Chicago Press, 2021); Matthew Maycock, 'They're All Up in the Gym and All That, Tops Off, Fake Tan': Embodied Masculinities, Bodywork and Resistance Within Two British Prisons', in *New Perspectives on Prison Masculinities*, edited by Matthew Maycock and Kate Hunt, pp. 65–89 (London: Palgrave Macmillan, 2018); Faye Woods, 'Structured Reality: Designer Clothes, Fake Tans, Real Drama?', in *British Youth Television: Transnational Teens, Industry, Genre*, pp. 185–221 (London: Palgrave Macmillan, 2016); Shirley Anne Tate, ' "The Browning", Straighteners, and Fake Tan', in *Black Beauty, Aesthetics, Stylization, Politics*, pp. 99–122 (London and New York: Routledge, 2016).

27 John Emsley, *Vanity, Vitality, and Virility: The Science behind the Products You Love to Buy* (Oxford: Oxford University Press, 2004), pp. 25–7; Steven Farmer, *Strange Chemistry: The Stories Your Chemistry Teacher Wouldn't Tell You* (New Jersey: John Wiley & Sons, 2017), p. 143.

28 Bill Cormack, *A History of Holidays, 1812–1990* (London: Routledge and Thomas Cook Archives, 1998), pp. 108–9.

29 To note changes in skin colour beauty trends, I quantified the different skin colour types of everyone on the front covers of the *Vogue and Gentlemen's Quarterly* magazines, per year, from the 1950s to the 2000s.

30 Anon., 'A Lotion That Tans.' *The Chemist and Druggist*, 18 June 1960, p. 726.

31 Anon., 'Sun and Sunless Tans.' *The Chemist and Druggist*, 2 July 1960, p. 5; in America, Coppertone's 'Quick Tan' (QT) was the leading but also short-lived DHA tanning product. Followed by 'Positan' – Farmer, *Strange Chemistry*, p. 143; Sherry Pagoto, 'Sunless Tanning', in *Shedding Light on Indoor Tanning*, edited by Carolyn Heckman and Sharon Manne, p. 166 (Dordrecht, Heidelberg, New York, and London: Springer, 2012); Anon., 'Artificial Suntan Preparations.' *Which?* December 1960, pp. 282–4.

32 Launched by Michael Young in 1957, this magazine's aim was to offer 'impartial, independent and scientifically-grounded factual information in order to promote rational choice in consumption'. Anon., '60 Years of Questioning. Which?' *Which?*, 2017. http://explore.which.co.uk/timeline. Accessed 22 October 2019; Anon., 'Artificial Suntan Preparations', pp. 282–4.

33 Anon., 'Artificial Suntan Preparations', pp. 282–4; Anon., 'Soap.' *Perfumery & Cosmetics* 33, no. 9 (September 1960): 957.

34 Anon., 'Artificial Suntan Preparations', pp. 282–4.

35 Anon., 'Artificial Suntan Preparations', p. 804.

36 Anon., 'Artificial Suntan Preparations', p. 282.

37 Anon., 'Living in the Bronze Age.' *Financial Times*, 24 July 1962, p. 8.

38 Audrey Baker and Gloria Baptist-Smith, 'WHO NEEDS SUNSHINE?' *She*, August 1970, p. 60; Cyril Ashley (Golden, Ltd), 'Overnight Tan Products Then and Now.' *The Chemist and Druggist*, 12 June 1971, pp. 4–5.

39 Anon., 'Tanfastic.' *She*, July 1972, p. 12.

40 Ashley, 'Overnight Tan Products Then and Now', pp. 4–5.

41 I could not find safety testing results for DHA in either Britain or America until the early 1970s. Pagoto, 'Sunless Tanning', pp. 165–78.

42 Caroline Richards, 'Summer's Burning Question: Will You Be Earthly Bronzed or Provocatively Pale?' *Cosmopolitan*, June 1974, p. 174.

43 Anon., 'ASK OUR EXPERTS ANYTHING COMPANY COUNSEL.' *Company*, May 1981, p. 131.

44 Ashley, 'Overnight Tan Products Then and Now', pp. 4–5; Anon., 'Piz Buin Two.' *The Chemist and Druggist*, 14 April 1979, p. 518; Anon., 'Vichy.' *The Chemist and Druggist*, 18 January 1986, p. 103.

45 Alanah Reid, 'A History of the Daily Mail.' Historic Newspapers, 17 September 2020. www.historic-newspapers.co.uk/blog/daily-mail-history/. Accessed 28 October 2020.

46 Adrian Bingham and Martin Conboy, *Tabloid Century: The Popular Press in Britain, 1896 to the Present* (Oxford: Peter Lang, 2015).

47 By 2010, if not before, the *Daily Mail* was the only newspaper to have a larger male than female readership. Carolyn M. Byerly,

The Palgrave International Handbook of Women and Journalism (London: Palgrave Macmillan, 2013), p. 178.

48 Miranda Ingram, 'Our Beauties of the Boudoir. An Image to Cherish.' *Daily Mail*, 18 May 1989, p. 13.

49 Newby Hands, 'Tanning without the Tears. How to Keep That Golden Glow All Year Round.' *Daily Mail*, 30 May 1991, p. 23.

50 Anon., 'The Safe-Tan Plan.' *Ladies' Home Journal*, 5 May 1992, p. 46.

51 Anon., 'Clarins Go Brown for Winter.' *The Chemist and Druggist*, 10 November 1990, p. 830; £9.75 in 1990 was the equivalent value of £23.27 in 2021 – CPI Inflation Calculator. www.officialdata.org/uk/inflation/1990?amount=9.75. Accessed 8 December 2021.

52 Hands, 'Tanning without the Tears', p. 23.

53 Hands, 'Tanning without the Tears', p. 23.

54 Hands, 'Tanning without the Tears', p. 23.

55 Hands, 'Tanning without the Tears', p. 23.

56 Louise Atkinson, 'Day Two of Our Ten-Day Quick Weight-Loss Plan, a Healthy Tan and Exercises for Thighs. Protect your Skin with a New Tan.' *Daily Mail*, 13 July 1993, pp. 42–3.

57 Hayward, 'A Golden Tan – and No Streaks', p. 36.

58 Hayward, 'A Golden Tan – and No Streaks', p. 36.

59 Hayward, 'A Golden Tan – and No Streaks', p. 36.

60 Hayward, 'A Golden Tan – and No Streaks', p. 36.

61 Atkinson, 'Day Two of Our Ten-Day Quick Weight-Loss Plan', pp. 42–3.

62 Sarah Clarke, 'THE COMPLETE GUIDE TO FAKING IT.' *She*, June 1993, pp. 142–4; Anon., 'New Ways to Fake a Tan.' *Company*, July 1993, p. 13; Roger Tredre, 'Get Real and Get a Fake Tan This Year.' *The Guardian*, 4 June 1995, p. 7.

63 Newby Hands, 'The Good Fakes Guide.' *Daily Mail*, 21 May 1992, p. 31; Christina Probert Jones, 'Bronze Age.' *Observer*, 16 May 1993, p. B54; Michele Goldsmith, 'The Fakes' Progress.' *Daily Mail*, 30 July 1994, p. 45.

64 Judy Rumbold, 'Tans Harder to Fake than an Orgasm.' *Observer*, 17 May 1998, p. 79.

65 Aidin, 'When Faking It Doesn't Cause a Stink', p. 37.

66 Aidin, 'When Faking It Doesn't Cause a Stink', p. 37.

67 Newby Hands, 'Men, Make-Up and Machismo.' *Daily Mail*, 7 April 1992, p. 13.

68 Louise Chunn, 'A Generous Cut above the Rest.' *The Guardian*, 8 June 1992, p. 22; Imogen Edwards-Jones, 'Browned Off with the Tanning Business.' *The Guardian*, 24 May 1995, p. A8.

69 Louise Atkinson, 'GOOD HEALTH. Tomorrow's Tan. Can an Injection Really Give You a Suntan?' *Daily Mail*, 22 June 1993, p. 35.

70 Atkinson, 'GOOD HEALTH', p. 35.

71 Hadley's research on A-MSH led to biological, biochemical, pharmacological, endocrinological, physiological, and medical investigations. This research led to the discovery, and later development, of both MelanoTan I and MelanoTan II. Victor J. Hruby, 'Professor Mac E. Hadley.' *General and Comparative Endocrinology* 151, no. 3 (2007): 358–60; B. Ramos-Molina, M.G. Martin, and I Lindberg, 'PCSK1 Variants and Human Obesity', in *Genetics of Monogenic and Syndromic Obesity*, edited by Ya-Xiong Tao, pp. 47–74 (London: Academic Press, 2016).

72 Atkinson, 'GOOD HEALTH', p. 35.

73 Atkinson, 'GOOD HEALTH', p. 35.

74 Patrick Riley, 'Carcinogenesis: When transmission of epigenetic information goes awry', *Health and Medicine Research OUTREACH*, 105 (2018).

75 Atkinson, 'GOOD HEALTH', p. 35.

76 Atkinson, 'GOOD HEALTH', p. 35.

77 James Mills, 'Can One Drug Really Give You a Tan, Make You Slim and Boost Your Sex Life?' *Daily Mail*, 12 April 2004, p. 22.

78 Anon., 'Medicines Agency Warns on Unlicensed Tanning Drugs.' *The Chemist and Druggist*, 22 November 2008, p. 25.

79 Anon., 'Medicines Agency Warns on Unlicensed Tanning Drugs.'

80 L. Habbema, A.B. Halk, M. Neumann, and W. Bergman, 'Risks of Unregulated Use of Alpha-Melanocyte-Stimulating Hormone Analogues: A Review.' *International Journal of Dermatology* 56 (2017): 975–80.

81 See note 14.

82 British Film Institute Stephen Street Archive, London, 'Tanorexia.' *Esther*, BBC2, 18 June 1997.

83 Anon., 'Franchises. START YOUR OWN BUSINESS.' *Daily Mail*, 7 June 1993, p. 36.

84 Wellcome Trust Collection, London, Box 628, Sunbathing Ephemera; Box 1, The Tanning Shop Voucher, 29 July 1994; Hayward, 'A Golden Tan – and No Streaks', and Sills, 'The Six-Minute Tanning Chamber', p. 35.

85 Steve Tooze, 'Tanorexia.' *Daily Mail*, 16 May 1996, pp. 44–5.

86 Penny Tinkler, *Smoke Signals: Women, Smoking and Visual Culture in Britain* (Oxford: Berg, 2006).

87 Ian Murray (Medical Correspondent), 'Sunbed Clients May Be Paying with Their Lives.' *The Times*, 3 March 1999, p. 6; the

sample base of this study consisted of 6,143 adults. Taylor Nelson, 'Consumer Research on Sun Tanning and Sunbeds', *The Sunbed Association*, 1997.
88 See note 26.
89 Nina Jablonski, *Skin: A Natural History* (Berkeley, CA, Los Angeles, CA, and London: University of California Press, 2013), p. 185.

Bibliography

Aidin, Beatrice. 'When Faking It Doesn't Cause a Stink.' *Financial Times*, 1 August 2009.
Anonymous. '60 Years of Questioning. Which?' *Which?*, 2017. http://explore. which.co.uk/timeline. Accessed 22 October 2019.
Anonymous. 'Artificial Suntan Preparations. A Report on Consumer Tests.' *The Chemist and Druggist*, 31 December 1960.
Anonymous. 'ASK OUR EXPERTS ANYTHING COMPANY COUNSEL.' *Company*, May 1981.
Anonymous. 'Clarins Go Brown for Winter.' *The Chemist and Druggist*, 10 November 1990.
Anonymous. 'COMPANY COUNSEL.' *Company*, October 1985.
Anonymous. *Domestic Appliances and Personal Care Products*. Philips English Catalogue. Eindhoven: Philips International Company Archive, February 1997.
Anonymous. 'Franchises. START YOUR OWN BUSINESS.' *Daily Mail* , 7 June 1993.
Anonymous. 'Headnotes!' British Pathe Archive, 1939. www.britishpathe. com/video/headnotes/query/fake+tan. Accessed 9 December 2021.
Anonymous. 'A Lotion That Tans.' *The Chemist and Druggist*, 18 June 1960.
Anonymous. 'Living in the Bronze Age.' *Financial Times* , 24 July 1962.
Anonymous. 'Medicines Agency Warns on Unlicensed Tanning Drugs.' *The Chemist and Druggist*, 22 November 2008.
Anonymous. 'New Ways to Fake a Tan.' *Company*, July 1993.
Anonymous. 'No Need to Ever Get Sunburnt. Golden Tan. Completely Safe and without the Sun.' *Daily Mail*, 4 June 1992.
Anonymous. 'Piz Buin Two.' *The Chemist and Druggist*, 14 April 1979.
Anonymous. 'The Safe-Tan Plan.' *Ladies' Home Journal*, 5 May 1992.
Anonymous. 'Soap.' *Perfumery & Cosmetics* 33, no. 9 (September 1960): 957.
Anonymous. 'Sun and Sunless Tans.' *The Chemist and Druggist*, 2 July 1960.
Anonymous. 'Tanfastic .' *She* , July 1972.
Anonymous. 'TAN NATURALLY WITHOUT THE SUN.' *Daily Mail* , 20 April 1991.
Anonymous. 'Vichy.' *The Chemist and Druggist*, 18 January 1986.

Ashley, Cyril. 'Overnight Tan Products Then and Now.' *The Chemist and Druggist*, 12 June 1971.

Atkinson, Louise. 'Day Two of Our Ten-Day Quick Weight-Loss Plan, a Healthy Tan and Exercises for Thighs. Protect your Skin with a New Tan.' *Daily Mail*, 13 July 1993.

Atkinson, Louise. 'GOOD HEALTH. Tomorrow's Tan. Can an Injection Really Give You a Suntan?' *Daily Mail*, 22 June 1993.

Baker, Audrey, and Gloria Baptist-Smith. 'WHO NEEDS SUNSHINE?' *She*, August 1970.

Berridge, Virginia. 'Electronic Cigarettes and History.' *Lancet* 383, no. 9936 (2014): 2204–5.

Berridge, Virginia, and Penny Starns. 'The "Invisible Industrialist" and Public Health: The Rise and Fall of "Safer Smoking" in the 1970s', in *Medicine, the Market and Mass Media: Producing Health in the Twentieth Century*, edited by Virginia Berridge and Kelly Loughlin, pp. 172–91. Abingdon and New York: Routledge, 2012.

Bingham, Adrian, and Martin Conboy. *Tabloid Century: The Popular Press in Britain, 1896 to the Present*. Oxford: Peter Lang, 2015.

British Film Institute Stephen Street Archive, London. 'Tanorexia.' *Esther*, BBC2, 18 June 1997.

Bulman, A. 'Letters. People Are Overusing Sunbeds.' *British Medical Journal* 310, no. 6990 (1995): 1327.

Byerly, Carolyn M. *The Palgrave International Handbook of Women and Journalism*. London: Palgrave Macmillan, 2013.

Cameron, I.H., and Christine McGuire. 'Are You Dying to Get a Suntan?' – Pre- and Post-Campaign Survey Results.' *Health Education Journal* 49, no. 4 (1990): 166–70.

Carter, Simon. *Rise and Shine: Sunlight, Technology and Health*. Oxford: Berg, 2007.

Chunn, Louise. 'A Generous Cut above the Rest.' *The Guardian*, 8 June 1992.

Clarke, Sarah. 'THE COMPLETE GUIDE TO FAKING IT.' *She*, June 1993.

Cormack, Bill. *A History of Holidays, 1812–1990*. London: Routledge and Thomas Cook Archives, 1998.

CPI Inflation Calculator. www.officialdata.org/uk/inflation/1990?amount= 9.75. Accessed 8 December 2021.

Creed, Fabiola. 'From "Immoral" Users to "Sunbed Addicts": The Media-Medical Pathologising of Working-Class Consumers and Young Women in Late Twentieth-Century England.' *Social History of Medicine* 35, no. 3 (2022): 770–92.

Creed, Fabiola. *The Rise and Fall of the Sunbed in Britain: Tanning Culture from Fad to Fear*. London: Bloomsbury, 2025.

Davis, H.J., and P.P.S. (eds). *The Bulletin* 270, no. 270. Nottingham: Boots Archive, 21 March 1952.

Department of Health. *The Health of the Nation – A Strategy for Health in England*. London: Her Majesty's Stationery Office, 1992.

Deslandes, Paul. *The Culture of Male Beauty in Britain: From the First Photographs to David Beckham*. Chicago, IL, and London: University of Chicago Press, 2021.

Edgerton, David. *The Shock of the Old: Technology and Global History since 1900*. London: Profile Books, 2011.

Edwards-Jones, Imogen. 'Browned Off with the Tanning Business.' *The Guardian*, 24 May 1995.

Emsley, John. *Vanity, Vitality, and Virility: The Science behind the Products You Love to Buy*. Oxford: Oxford University Press, 2004.

Farmer, Steven. *Strange Chemistry: The Stories Your Chemistry Teacher Wouldn't Tell You*. New Jersey: John Wiley & Sons, 2017.

Freund, Daniel. *American Sunshine: Diseases of Darkness and the Quest for Natural Light*. Chicago, IL: University of Chicago Press, 2012.

Goldsmith, Michele. 'The Fakes' Progress.' *Daily Mail*, 30 July 1994.

Habbema, L., A.B. Halk, M. Neumann, and W. Bergman. 'Risks of Unregulated Use of Alpha-Melanocyte-Stimulating Hormone Analogues: A Review.' *International Journal of Dermatology* 56 (2017): 975–80.

Haiken, Elizabeth. *Venus Envy: A History of Cosmetic Surgery*. Baltimore, MD: Johns Hopkins University, 1997.

Hall, Ronald E. (ed.). *The Melanin Millennium: Skin Color as 21st Century International Discourse*. Dordrecht, Heidelberg, New York, and London: Springer, 2013.

Hands, Newby. 'The Good Fakes Guide.' *Daily Mail*, 21 May 1992.

Hands, Newby. 'Men, Make-Up and Machismo.' *Daily Mail*, 7 April 1992.

Hands, Newby. 'Tanning without the Tears. How to Keep That Golden Glow All Year Round.' *Daily Mail*, 30 May 1991.

Hansen, Devon. 'Shades of Change: Suntanning and the Twentieth-Century American Dream.' PhD dissertation, Boston University, 2007.

Hayward, Katie. 'A Golden Tan – and No Streaks.' *Daily Mail*, 6 May 1993.

Heckman, Carolyn, and Sharon Manne (eds). *Shedding Light on Indoor Tanning*. Dordrecht, Heidelberg, New York, and London: Springer, 2012.

Hope, Jenny. 'Fear Mounts as Sunbed Fans Ignore the Health Risks.' *Daily Mail*, 5 October 1990.

Hruby, Victor J. 'Professor Mac E. Hadley.' *General and Comparative Endocrinology* 151, no. 3 (2007): 358–60.

Hunter, David J., Naomi Fulop, and Morton Warner. *From 'Health of the Nation' to 'Our Healthier Nation'*. Copenhagen: World Health Organization, Regional Office for Europe, Policy Learning Curve Series, no. 2, August 2000.

Ingram, Miranda. 'Our Beauties of the Boudoir. An Image to Cherish.' *Daily Mail* , 18 May 1989.

Jablonski, Nina. *Living Color: The Biological and Social Meaning of Skin Color*. Berkeley, CA, Los Angeles, CA, and London: University of California Press, 2012.

Jablonski, Nina. *Skin: A Natural History*. Berkeley, CA, Los Angeles, CA, and London: University of California Press, 2013.

Jeffreys, Sheila. *Beauty and Misogyny: Harmful Cultural Practices in the West*. London: Routledge, 2005.

Maycock, Matthew, and Kate Hunt (eds). *New Perspectives on Prison Masculinities*. London: Palgrave Macmillan, 2018.

Mills, James. 'Can One Drug Really Give You a Tan, Make You Slim and Boost Your Sex Life?' *Daily Mail*, 12 April 2004.

Mold, Alex. *Making the Patient-Consumer: Patient Organisations and Health Consumerism in Britain*. Manchester: Manchester University Press, 2015.

Mold, Alex, Peder Clark, Gareth Millward, and Daisy Payling. *Placing the Public in Public Health in Post-War Britain, 1948–2012*. London: Palgrave Macmillan, 2019.

Murray, Ian. 'Sunbed Clients May Be Paying with Their Lives.' *The Times*, 3 March 1999.

Nelson, Taylor. 'Consumer Research on Sun Tanning and Sunbeds.' *The Sunbed Association*, 1997.

Office of Population Censuses and Surveys, Social Survey Division. *OPCS Omnibus Survey, UK Data Service, SN: 3739, January 1995*. 1998. http://doi.org/10.5255/UKDA-SN-3739–1. Accessed 25 January 2020.

Peiss, Kathy. *Hope in a Jar: The Making of America's Beauty Culture*. New York: Metropolitan Books, 1998.

Probert Jones, Christina. 'Bronze Age.' *Observer*, 16 May 1993.

Reid, Alanah. 'A History of the Daily Mail.' Historic Newspapers, 17 September 2020. www.historic-newspapers.co.uk/blog/daily-mail-hist ory/. Accessed 28 October 2020.

Richards, Caroline. 'Summer's Burning Question: Will You Be Earthily Bronzed or Provocatively Pale?' *Cosmopolitan*, June 1974.

Riley, Patrick. 'Carcinogenesis: When Transmission of Epigenetic Information Goes Awry.' *Health and Medicine Research OUTREACH*, 105, 2018.

Romano, Sally. 'The Dark Side of the Sun: Skin Cancer, Sunscreen, and Risk in Twentieth-Century America'. PhD dissertation, Yale University, 2006.

Rumbold, Judy. 'Tans Harder to Fake than an Orgasm.' *Observer*, 17 May 1998.

Schlich, Thomas, and Ulrich Tröhler (eds). *The Risks of Medical Innovation: Risk Perception and Assessment in Historical Context*. London and New York: Routledge, 2004.

Segrave, Kerry. *Suntanning in 20th Century America*. Jefferson, NC: McFarland & Company, Inc, 2005.

Severi, G., G.G. Giles, C. Robertson, P. Boyle, and P. Autier. 'Mortality from Cutaneous Melanoma: Evidence for Contrasting Trends between Populations.' *British Journal of Cancer* 82, no. 11 (2000): 1887–91.

Shuttleworth, David. 'Sunbeds and the Pursuit of the Year Round Tan Should Be Discouraged.' *British Medical Journal* 307 (1993): 1508–9.

Sills, Amanda. 'The Six-Minute Tanning Chamber.' *Daily Mail*, 6 May 1993.

Tao, Ya-Xiong (ed.). *Genetics of Monogenic and Syndromic Obesity*. London: Academic Press, 2016.

Tate, Shirley Anne. ' "The Browning", Straighteners, and Fake Tan', in *Black Beauty, Aesthetics, Stylization, Politics*, pp. 99–22. London and New York: Routledge, 2016.

Thomas, Lynn N. *Beneath the Surface: A Transnational History of Skin Lighteners*. Durham, NC: Duke University Press, 2020.

Tinkler, Penny. *Smoke Signals: Women, Smoking and Visual Culture in Britain*. Oxford: Berg, 2006.

Tomes, Nancy. *Remaking the American Patient: How Madison Avenue and Modern Medicine Turned Patients into Consumers*. Chapel Hill, NC: University North Carolina Press, 2016.

Tooze, Steve. 'Tanorexia.' *Daily Mail*, 16 May 1996.

Tredre, Roger. 'Get Real and Get a Fake Tan This Year.' *The Guardian*, 4 June 1995.

Van der Post, Lucia, and Joan Price. 'Browned Off without Sun.' *Financial Times*, 28 June 1980.

Wailoo, Keith. *How Cancer Crossed the Color Line*. Oxford: Oxford University Press, 2010.

Wellcome Trust Collection, London. Box 628, Sunbathing Ephemera; Box 1, The Tanning Shop Voucher, 29 July 1994.

Woloshyn, Tania. 'Le Pays du Soleil: The Art of Heliotherapy on the Côte d'Azur.' *Social History of Medicine* 26, no. 1 (2013): 74–93.

Woloshyn, Tania. *Soaking Up the Rays: Light Therapy and Visual Culture in Britain, c. 1890–1940*. Manchester: Manchester University Press, 2017.

Woods, Faye. *British Youth Television: Transnational Teens, Industry, Genre*. London: Palgrave Macmillan, 2016.

7

Against 'prevention pills': North American breast cancer activists and chemoprevention

Grazia De Michele

In a 2013 monograph on tamoxifen, pharmacologist and breast oncologist Craig Jordan and his co-authors described the story of the drug as one of unparalleled singularity. Destined to be hailed as a breakthrough in the treatment of breast cancer, tamoxifen was originally nothing more than a failed morning-after pill. However – they argued – thanks to the collaboration between industry and researchers, the lives of 'hundreds of thousands, perhaps millions, of women' were saved.[1] Tamoxifen was the first member of a new class of drugs now known as selective estrogen receptor modulators (SERMs) that act as both oestrogen agonists and antagonists depending on the organ. It was synthesized in 1962 by Dora Richardson, a chemist at the British corporation Imperial Chemical Industries, as part of a programme to develop oral contraceptives. This initiative was led by biologist Arthur Walpole, who also worked on the company's cancer programme. Owing to Walpole's dual interests, the patent filed for the new compound also covered the control of hormone-dependent cancers. In an attempt to understand its effect on cancer, a trial on menopausal and post-menopausal women with breast cancer was conducted at the turn of the 1970s at the Christie Hospital in Manchester. Results showed that, unlike other synthetic hormones used at the time to treat the disease, tamoxifen was very well tolerated and induced remarkable tumour regression in several patients. Trials conducted in Sweden also revealed that contrary to expectations, the drug did not work as a contraceptive since, even at low doses, it promoted ovulation rather than inhibiting it. Nevertheless, the market for a breast cancer drug was small due to the generally poor disease outcome, causing Imperial Chemical Industries to initially abandon the development of tamoxifen. This decision, however, was soon reversed because of internal pressures and concerns within the company that refusal to invest in a drug able to control cancer could undermine its reputation.[2]

However, a new use of tamoxifen emerged. By the 1980s a trial conducted in the United States by the National Cancer Institute proved that the use of tamoxifen significantly reduced the recurrence rate in women with oestrogen receptor-positive invasive breast cancer who had already undergone surgery. Preliminary results from another trial sponsored by the National Cancer Institute and involving women treated with lumpectomy and radiation for ductal carcinoma in situ (DCIS) – a non-invasive form of breast cancer – suggested that tamoxifen decreased the risk of developing invasive cancer in the same or contralateral breast. Building on these findings in 1992, the National Cancer Institute started the so-called Breast Cancer Prevention Trial, a double-blind randomized controlled trial of tamoxifen versus a placebo, enrolling healthy women in over 324 sites in the United States and Canada. In 1998, tamoxifen was granted approval by the United States Food and Drug Administration (FDA), under the name Nolvadex, for use in women considered at high risk of breast cancer.[3]

Since the post-war period, pharmaceuticals – as historian Jeremy Greene has pointed out – have served as 'technolog[ies] of control, reshaping the formerly unruly contours of disease into forms more acceptable to human life and livelihood'.[4] Tamoxifen, for example, arguably contributed to turn breast cancer from a *treatable* into a *preventable* disease – just as breast cancer, in a sense, transformed a failed contraceptive into a cancer drug. The drug was hence not a mere tool but a proper actor involved in a web of processes and relationships, including the contention over its serious side effects: blood clots, cataracts, liver cancer in rats, and endometrial adenocarcinoma in women. On one side of the debate were those promoting tamoxifen –the manufacturer, the National Cancer Institute, and the US FDA – and on the other, groups of patient consumers eventually forming a coalition called Prevention First (PF) who just as vehemently rejected it.[5] Led by the San Francisco-based grassroots group Breast Cancer Action (BCA), PF included organizations that had been key players in the United States women's health movement. The coalition opposed the marketing of tamoxifen as preventative and chemopreventative therapeutics more generally. To this end, PF also proposed its own notion of prevention based on the Precautionary Principle, which stipulated that an activity or technology should be avoided until its safety has been proved, with the burden of proof falling on its proponents.[6] BCA as well as PF thus saw tamoxifen as part of a system in which cancer resulted from the exposure to toxic substances produced by the same for-profit entities that, in turn, manufactured drugs to treat and prevent the disease they had contributed to cause in the first place – drugs that had, ironically, a pathogenic effect themselves.

This chapter focuses on the rejection of tamoxifen as a preventative technology by BCA and the members of PF, as well as of the very idea of the pharmaceutical prevention of cancer. This historical episode has already been analysed through a sociological lens, focusing on how the pharmaceutical industry, the regulatory agency, and activists positioned themselves in the process of reconfiguring tamoxifen as a treatment for women at high risk of breast cancer.[7] However, using records from the hitherto unexplored archive of BCA, as well as existing literature on feminists' rejection of the use of synthetic hormones in women in the 1970s, I take a historical perspective, situating this episode within the broader history of the women's health movement. In doing so, I argue that the marketing of tamoxifen as a preventative technology is one more instance in the long history of selling artificial hormones to women – as, for example, in the cases of contraceptive pills with a high content of oestrogen and of the man-made oestrogen diethylstilboestrol – without in-depth knowledge of their long-term consequences or evidence of real risk. This practice, in the late 1990s as well as in the 1970s, was opposed by feminist patient consumers who held the pharmaceutical industry and the United States FDA accountable and mobilized women by providing them with unbiased information. Though usually presumed to be predominantly White and middle class, many of these health feminists were also Jewish and of vulnerable socioeconomic backgrounds. For these women, their experiences of marginality contributed to making them particularly sensitive to social injustice and willing to mobilize.[8]

Tamoxifen in a changed breast cancer and patient consumer landscape

Historians have argued that, in the nineteenth century, cancer, previously conceived of as a constitutional malady, was 'reconstructed as a "local" disease'.[9] Its treatment was thereby monopolized by surgeons, who, in the case of breast cancer, practised disfiguring and disabling mastectomies.[10] The last thirty years of the twentieth century, however, witnessed a change of paradigm or, according to sociologist Maren Klawiter, a change of 'disease regime'.[11] In the new 'regime of biomedicalization', breast cancer was reconfigured as a systemic disease through the introduction of chemotherapy and tamoxifen, diagnostical biopsy and surgery were separated, and mammographic screening was introduced.[12]

Systemic treatment prolonged the illness experience, multiplied the side effects, increased the number of affected body areas, and grew the number of specialists to be consulted. In addition, the separation of diagnosis from surgery enabled women to participate in the decision-making process regarding surgical techniques, especially once modified radical mastectomy and lumpectomy followed by radiation found their way into the clinical practice. The combination of empowerment and the totalizing state of sickness owing to a progressively longer course of therapy favoured the creation of support groups. These groups provided their members with the opportunity to become visible to each other, exchange information, establish bonds, and assume a collective identity centred on breast cancer. As a result, support groups became a breeding ground for organizations like BCA that, in the 1990s, were started with the aim of putting breast cancer at the forefront of the political arena.[13]

As Robert Aronowitz has highlighted, mammographic screening in this era turned the experience of breast cancer, once limited only to women with symptomatic disease, into 'a mass phenomenon' involving a wider group of subjects considered at risk of developing it.[14] Prominent figures scientifically supported the idea that the extent of the breast cancer problem was much bigger than previously thought. In 1992 Bernard Fisher, chair of the National Cancer Institute's clinical trials cooperative group testing tamoxifen and regarded as one of the leading breast cancer experts of the twentieth century, argued that women diagnosed with breast cancer through clinical examination or mammography 'compris[ed] only one cohort of the female population whose breasts contain a spectrum of aberrations'.[15] The other cohorts consisted of women with lesions detectable exclusively by mammography; women with lesions that could not be detected but, if malignant, might be in the future; women whose breasts, albeit apparently normal, might have been subject to 'biological alterations' that could lead to breast cancer;[16] and women with no lesions or biological changes but still 'at risk for developing breast cancer even though they [might] have no definable risk factors'.[17] It was thus necessary to devise a strategy to prevent the progression of undetectable breast cancers. Among the several agents worthy of consideration, tamoxifen was – in Fisher's view – the most promising. Its efficacy in reducing the risk of recurrence in women with invasive breast cancer and of invasive breast cancer in those with DCIS had indeed already been demonstrated and its side effects deemed minimal.

The approach to prevention based on early detection and therapeutics for breast cancer was not a novelty. Since the early twentieth century, especially in the United States, cancer prevention – as David Cantor has pointed out – came to be framed as early detection of further growth of already existing cancers or of precancerous conditions followed by surgery and, later, radiotherapy. The 1960s and 1970s witnessed a '*reinvention*' of an older tradition of interest in the environmental and lifestyle causes of cancer anchored to statistics and risk calculation.[18] By the early 1980s, though, the barycentre of the scientific discourse on cancer had moved back to therapeutic prevention. In their report on the 'Causes of Cancer', commissioned by the Congress's Office for Technology Assessment, epidemiologists Richard Doll and Richard Peto noted that 'protective agents [might] be of more practical importance than causative agents for it [might] be easier to prescribe than to proscribe'.[19] In those years, cancer chemoprevention – the pharmacological prevention of cancer – was taking its first steps. Pharmacologist Michael Sporn, one of its pioneers, deemed cancer a process starting well before the disease was detectable. The progression of precancerous lesions to a full-blown malignant state could thus be stopped or even reversed with the employment of synthetic chemical agents.[20]

Within a decade, the field of chemoprevention became one of the top research priorities of the National Cancer Institute.[21]

The twentieth century, however, was also characterized by its share of public criticism of medical technology, including pharmaceuticals. Such critiques were frequently made by patients, a fact that helped fuel the rise of patient advocacy and consumerism as a movement. Nancy Tomes is among those historians to challenge the assumption that the phenomenon was peculiar to the 1970s. Nevertheless, it was in the late 1960s that patient consumers adopted a more antagonistic attitude towards what they saw as instruments of power and oppression within a White, androcentric, and capitalist medical system. Science and medicine were portrayed as part of the medical-industrial complex, a phrase coined by the Health Policy Advisory Center collective that used it in *The American Health Empire*, published in 1970. Furthermore, unlike their predecessors, 1970s patient consumers believed that all the individuals, regardless of their class, race, sex, and health state, knew their bodies like anyone else and were entitled to make choices about them.[22]

Sharing these critiques, concerns over healthcare in the reproductive sphere in feminist consciousness-raising groups led to the emergence of the women's health movement in the late 1960s. Within the movement, synthetic hormones quickly became a major source of worry. In 1969 the publication of Barbara Seaman's *The Doctors' Case against the Pill* raised alarm about the dangers of contraceptive birth control pills and the extent to which women were not being made aware of them.[23] When Seaman and the women who had been harmed by the pill were not invited to testify at the Congressional hearings on the subject, called by Senator Gaylord Nelson, protests and counter-hearings were organized by the feminist group, D.C. Women Liberation, led by Alice Wolfson. During those same years, Belita Cowan, a master's student at the University of Michigan and a part-time worker at the University Hospital in Ann Arbor, started Advocates for Medical Information (AMI) together with other women. The goal of the organization was to raise awareness about the dangers of diethylstilboestrol. The compound had been given to students as a morning-after pill within a study conducted at the university at the same time that reports about its links to vaginal cancer in the daughters of pregnant women who ingested the drug appeared in the medical literature. The AMI hugely contributed to the ensuing campaign to expose its dangers. In December 1975 a memorial service for all the victims of diethylstilboestrol, the pill, and synthetic oestrogens took place on the steps of the FDA in Maryland, marking the birth of the National Women's Health Network (NWHN). Its founders, Barbara Seaman, Alice Wolfson, Belita Cowan, Phyllis Chesler, and Mary Howell, established that the organization would never accept money from pharmaceutical companies.[24]

The NWHN was among the first organizations that the founders of BCA – Elenore Pred, Susan Claymon, Linda Reyes, and Belle Shayer – contacted when they established their organization in 1990. Pred, a veteran of the civil rights and the feminist movements, had met Claymon and Reyes through a support group for women living with metastatic breast cancer in San Francisco, while she and Shayer had become friends after attending a retreat for people with different forms of cancer organized by a centre specializing in holistic healing.[25] The four women invited others to join them at Pred's home on 1 July 1990 to 'organize Breast Cancer Action'.[26] As they stated in the flyer announcing their first meeting: 'Our goals are education and political action to prevent a further rise in the incidence of breast cancer; indeed we hope that our efforts will serve in the future to lower the breast cancer rate in the United States.'[27] The meeting was also attended by healthy women, and not only in solidarity; their presence reflected in part a changing perception of breast cancer as a disease that every woman was at risk of developing.

Tamoxifen for the chemoprevention of breast cancer and its adversaries

When the launch of the Breast Cancer Prevention Trial was announced by the National Cancer Institute in 1991, the NWHN declared its opposition to the trial for a variety of reasons. Manufacturers of tamoxifen – the NWHN maintained – had reported 'high incidence of liver tumors and cancers in rats given doses similar to those proposed for prevention and … possible higher risk of liver cancer in breast cancer patients'.[28] Furthermore, in healthy women, the drug could trigger 'chemically induced menopause, which can cause uncomfortable changes even in women who experienced natural menopause years earlier'.[29] The National Cancer Institute was planning to enrol '16,000 high risk women ... in a five-year trial [whereas] only 2,000 women with breast cancer had used tamoxifen in controlled trials for five years'.[30] The Network hence asked to wait until more women with breast cancer had used the drug for a longer period, thereby providing all the stakeholders with more evidence on which the decision for a trial on healthy subjects could be based. The organization indeed did not oppose the use of tamoxifen for the treatment of women diagnosed with breast cancer. It was the administration of a drug with known and suspected high toxicity to *healthy* women that was regarded as unacceptable. In an article published in *The Lancet* and significantly titled 'Tamoxifen: Disease Prevention or Disease Substitution?', Network member Adriane Fugh-Berman concluded that 'although less toxic than conventional chemotherapy, [the compound] fail[ed] the more stringent standard of safety that is imperative for a primary prevention measure directed at the healthy general population'.[31]

Champions of cancer chemoprevention, such as pharmacologist Michael Sporn, judged these interpretations concerning the ethical implications of administering cancer drugs to healthy people before a diagnosis of cancer was made as 'fallacies'.[32] Cancer – he wrote in 1991 – was 'an endemic pestilence' still claiming too many lives.[33] The unwillingness on the part of scientists to recognize it as 'a process' of which 'everyone is part' had led them to focus all their efforts on treatment rather than on stopping carcinogenesis chemically.[34] As for breast cancer, given the efficacy of tamoxifen to reduce the risk of recurrence in women, and as a chemopreventative therapeutic in rats, Sporn expressed full support for the Breast Cancer Prevention Trial.

BCA shared the NWHN's take on the use of tamoxifen for the chemoprevention of breast cancer. Since its inception, the group had monitored the new evidence about the use of the drug and updated members through its widely circulated newsletter. A report on tamoxifen written by members of the organization was published in 1991. For some women, it concluded, tamoxifen had acted as a ' "miracle" drug, stopping tumor growth and allowing a period of disease-free time'.[35] For others, however, it had proved unsuccessful in 'preventing further metastasis'.[36] The report also offered an overview of the reasons why the NWHN opposed the idea of using 'tamoxifen as a breast cancer preventative, allowing "healthy" women at high risk to take the drug'.[37]

BCA collaborated closely with the NWHN to try to stop or at least convince the National Cancer Institute to postpone the Breast Cancer Prevention Trial. In spite of their efforts, the trial began in April 1992. From the beginning, the trial was plagued by scandals, affecting mainly the clinical trials cooperative group in charge of conducting it. Simultaneously, new evidence about the side effects of tamoxifen emerged, causing the International Agency for Research on Cancer and the State of California to list the drug as a carcinogen.[38] In the meantime, BCA not only kept informing its members on the Breast Cancer Prevention Trial and testified in public hearings concerning the carcinogenicity of tamoxifen, but campaigned on the connections between environmental pollution, the pharmaceutical industry, and breast cancer, with a particular focus on the company producing the drug. In October 1993, then President Nancy Evans penned an article on National Breast Cancer Awareness Month bringing attention to information she deemed relevant for readers. Tamoxifen's manufacturer, she said, was 'one of the co-founders and major funder' of National Breast Cancer Awareness Month – that she viewed as an advertising rather than an awareness campaign.[39] Moreover, the company not only manufactured tamoxifen, sold under the brand name of Nolvadex, but also 'organochlorines, chlorine-based industrial chemicals whose end products [could] range from DDT [dichlorodiphenyltrichloroethane], PCBs [polychlorinated biphenyls], Agent Orange' that were associated with 'increased incidence of breast cancer'.[40] Interestingly enough, Evans noted, the word 'environment' was never mentioned in the educational material produced for National Breast Cancer Awareness Month. Based on these facts, Evans concluded that '[National Breast Cancer Awareness Month] = more awareness = more mammograms = increased diagnosis of breast cancer = more prescriptions written for Nolvadex/tamoxifen' and provocatively questioned if National Breast Cancer Awareness Month could also be called 'Breast Cancer: A Growth Industry?'.[41]

In 1994, BCA was also one of the founding organizations of the Toxic Link Coalition (TLC), gathering environmental and health groups in the Bay Area. The aim of the coalition was to expose the responsibility of certain corporations for the levels of environmental pollution harming the health of people living in the area. Every October, during National Breast Cancer Awareness Month – which activists had sarcastically renamed National Cancer Industry Month – the TLC organized the so-called Toxic Tour of San Francisco's financial district, stopping outside the headquarters of corporate polluters, such as Chevron, staging protests and chanting 'Stop Cancer Where It Starts'.[42] Cohering with its anti-corporate stance, in 1998, BCA – by then led by former civil rights attorney Barbara Brenner – adopted the policy of not accepting donations from corporate polluters and the pharmaceutical industry. This decision, made after years of debate inside and outside of the organization, put BCA in the position to claim the role as watchdog of the breast cancer movement.[43]

Against 'prevention pills'

In April 1998 the Breast Cancer Prevention Trial was suddenly ended, fourteen months ahead of schedule, 'because of the clear evidence that tamoxifen reduced breast cancer risk'.[44] Even so, tamoxifen, the press release specified, had increased the risk of 'three rare but life-threatening health problems': endometrial cancer, pulmonary embolism, and deep-vein thrombosis.[45] In the June/July 1998 issue of BCA's newsletter, Brenner explained what the organization deemed to be wrong with the trial. While acknowledging that 'tamoxifen work[ed] – both as a treatment for metastatic disease and to reduce the risk of recurrence, that is, occurrence in the "healthy" breast – for some women', she pointed out that the drug was a 'known carcinogen [since] it increas[ed] the risk of endometrial cancer [and had] other serious potential side effects'.[46] After discussing the damage to medical knowledge involved in stopping the trial before its scheduled end, she highlighted how it was in the 'interest of all women and men [to] insist that the National Cancer Institute distribute the complete scientific data from the abbreviated [Breast Cancer Prevention Trial] as soon and as widely as possible'. This was necessary in order for the data to be analysed and to make sure that the public could understand 'what we know and what we don't as a result of the trial'.[47] Besides the right to receive proper information before deciding whether or not to use tamoxifen, Brenner's main point was to invite her readers to 'ask again and again why it is that when we demand true prevention of breast cancer, the answer we are offered is a pill'.[48] Such an invitation was made even more urgent by the news, reported in the media as the newsletter was going to press, that raloxifene, a new drug for osteoporosis, could soon replace tamoxifen for breast cancer prevention because it did not increase the risk of endometrial cancer.[49]

Nevertheless, by the end of October of that year, tamoxifen was granted approval by the FDA for the purpose of reducing 'the incidence of breast cancer in women at high risk for breast cancer'.[50] Initially, Zeneca – the new pharmaceutical company created in 1993 by Imperial Chemical Industries, which, in 1999, became AstraZeneca – had asked the FDA to allow the marketing of Nolvadex 'as a preventive agent for the reduction of breast cancer'.[51] The company had specified that 'the term prevention indicate[d] reduction in incidence or risk of invasive breast cancer over the period of time of the [trial] and [did] not necessarily imply that the initiation of breast cancer ha[d] been prevented or that the tumors ha[d] been permanently eliminated'.[52] BCA labelled such a definition of prevention 'Orwellian', urging the FDA, together with other women's health organizations, to reject it.[53] Furthermore, in a letter to the then director of the National Cancer Institute Richard Klausner, BCA pointed at the 'clear conflict of interest' represented by the support to 'efforts to broadly distribute tamoxifen as "prevention" to large numbers of healthy women' by 'at least one major breast cancer advocacy group' accepting 'large donations' from the company.[54]

After getting approval for the new indication for tamoxifen, Zeneca started an aggressive direct-to-consumer advertising campaign to promote it, favoured by new guidelines for TV advertising issued by the FDA in 1997.[55] Its 'Myths' commercial, without mentioning the name of the drug or its manufacturer, showed several women of different ages and races engaged in their daily activities. Their voices could be heard as they were making self-reassuring statements about their low risk of developing breast cancer, such as 'it's not in my family' and 'breast cancer is caused by a gene and we don't have it', while superimposed text showed data proving them wrong.[56] Then, a voice-over informed viewers that 'many women at risk for breast cancer don't know it'.[57] However, by calling the shown toll-free number, it was possible to obtain more information about the chances of developing the disease and how to reduce them. 'Why would I want to know if I'm at risk if there's nothing I can do?' a woman asked.[58] 'There *is* something you can do', the voice-over answered firmly while the same sentence was superimposed on screen.[59] Again, the voice-over urged viewers to call the toll-free number since 'today your doctor has ways that may reduce your chances of getting breast cancer if you find you're at high risk [and] high risk or not, every woman owes it to herself to find out'.[60] In a memorandum sent to other activists, such as organizers of DES Action, the Boston Women's Health Book Collective, the NWHN, and the Massachusetts Breast Cancer Coalition, Barbara Brenner wrote that the commercial was 'so benign' that BCA had received calls from women asking for the toll-free number to call and who thought that they would receive information about diet and exercise.[61] Yet – according to Brenner – it was also 'dangerous'. For one, it abundantly featured women of Asian and African American heritage, while the vast majority of those who had taken part in the trial leading to the approval of tamoxifen for use in healthy women were White.[62] Second, it aimed to induce 'anxiety' by proving wrong the above mentioned self assuring statements about breast cancer that help women cope with the fear of developing the disease.[63] Third, its 'unstated subject' was tamoxifen, but viewers were not informed about the dangerous side effects of the drug.[64] Overall, it was 'an example of direct marketing' aiming to convince women to request access to tamoxifen without needing it and in the absence of any information about its side effects.[65] Women's requests were probably going to be easily satisfied. Brenner also added two anecdotes reflecting the enthusiasm of physicians towards the drug: an oncologist from Texas had told her that many of his colleagues, after attending a conference, had become convinced that tamoxifen could be prescribed to all women; a woman volunteering for BCA had asked her oncologist if he was prescribing tamoxifen to healthy women, receiving as an answer an emphatic 'oh, yes!'[66] The commercial was indeed a success. According to the *Journal of the*

National Cancer Institute, by December 2000 it generated more than 700,000 calls to the toll-free number from people interested in receiving more information.[67]

After about two years of preparatory work, BCA, the organizations mentioned above, and other groups, such as the Center for Medical Consumers, the Women's Community Cancer Project, and Breast Cancer Action Montreal, formed the PF coalition. Their aim was to oppose the model of prevention based on 'drugs, medical products and procedures that may pose serious risk to human health' and demand 'true prevention of illness by reclaiming our right to safe water, air and food'.[68] The coalition planned to take a stand against direct-to-consumer advertising of drugs, to keep the public informed about the side effects and limits of 'prevention pills', to question the tendency to treat 'a risk factor as though it were a disease', to promote primary prevention, and to raise awareness of the 'Precautionary Principle'.[69] The latter was based 'on two commonsense ideas about protecting health and the environment'.[70] Its implementation 'require[d] that new drugs and technologies only be introduced into society when we have reasonably good evidence that they are safe, and that those who want to introduce a new drug or technology must first demonstrate that it is safe'.[71] To the members of PF, prevention thus meant contributing to a cleaner environment and protecting human health through the education and the mobilization of the public.

Conclusion

The second half of the twentieth century witnessed a major shift in the focus of medicine from alleviating the symptoms of diseases to addressing the risk of ill health through a vast array of screening and therapeutic technologies. As a result, the boundary between being at risk and experiencing a disease has become rather blurry. The advent of mammographic screening turned practically all women into at-risk subjects. At the same time, the introduction of systemic treatment made the illness experience much longer than it used to be for those who had received a diagnosis of breast cancer. Tamoxifen, a failed contraceptive, found its way into the United States market first as treatment for women who had been diagnosed with breast cancer and then as a preventative for healthy women deemed at high risk of developing it.

The rejection of tamoxifen for the chemoprevention of breast cancer and of chemoprevention tout court by BCA and the PF coalition is part of a longer history of clashes between a strictly medical approach to cancer prevention, privileging therapeutic technologies, and one emphasizing the environmental and social causes of the disease. Such rejection is also firmly rooted in patient consumerism and the women's health movement of the 1970s, particularly in its opposition to the medicalization and harm of women's bodies through the administration of synthetic hormones.

These continuities notwithstanding, some ruptures with respect to the past exist when it comes to the example of tamoxifen. BCA's focus on breast cancer is linked to the fragmentation and narrowing of a larger social movement and its transformation into a set of single-issue groups.[72] In turn, this phenomenon should be read against the backdrop of the almost uncontested and visible control or 'hegemony' – to use a term that Jean-Paul Gaudillière has invoked recently about synthetic hormones – by the industry, a hegemony that has been a distinguishing feature of Western medicine and society since the 1980s.[73] Over the last decades, disease has been turned into a 'simultaneously ... epidemiological event and a marketing event'.[74] Disease-based organizations, among which mainstream foundations and groups of breast cancer patient consumers constitute a model, have been recruited as major actors in this process.[75] BCA and PF are different, though. The work they carried out against the chemoprevention of breast cancer and promoting a drugless strategy to prevent it based on the Precautionary Principle aimed to ultimately benefit not only the patient consumers affected by this particular disease but the general population. Such work was made possible by the policy of not accepting funding from the pharmaceutical industry. The same decision had been taken by the NWHN in the 1970s. All this considered, BCA and PF can be placed with the minority of those who tried to urge women to resist the lure of medical technologies that – in the view of BCA and the organizations involved in the coalition – could put their health at risk.

Notes

1 Philipp Maximov, Russell Mcdaniel, and Craig Jordan, *Tamoxifen: Pioneering Medicine in Breast Cancer* (Basel: Springer Basel, 2013), p. xi.

2 Viviane Quirke, 'Imperial Chemical Industries and Craig Jordan, "the First Tamoxifen Consultant", 1960s–1990s.' *Ambix* 67, no. 3 (August 2020): 289–307. https://doi.org/10.1080/00026980.2020. 1794675.

3 Ilana Löwy, *Preventive Strikes: Women, Precancers and Prophylactic Surgery* (Baltimore, MD: Johns Hopkins University Press, 2010). On the so-called Gail model used for risk assessment, see Jennifer Fosket, 'Constructing "High-Risk Women": The Development and Standardization of a Breast Cancer Risk Assessment Tool.' *Science, Technology, & Human Values* 29, no. 3 (Summer 2004): 291–313. https://doi.org/10.1177/0162243904264960.

4 Jeremy Greene, *Prescribing by Numbers: Drugs and Definition of Disease* (Baltimore, MD: Johns Hopkins University Press, 2007), p. 226.

5 For an overview on drugs using the technology-in-practice approach, see Stefan Timmermans and Marc Berg, 'The Practice of Medical Technology.' *Sociology of Health and Illness* 25, no. 3 (April 2003): 97–114. https://doi.org/10.1111/1467-9566.00342.

6 Carolyn Raffensperger and Joel Tickner (eds), *Protecting Public Health and the Environment: Implementing the Precautionary Principle* (Washington, DC: Island Press, 1999).

7 Maren Klawiter, 'Risk, Prevention and the Breast Cancer Continuum: The NCI, the FDA, Health Activism and the Pharmaceutical Industry.' *History and Technology* 18, no. 4 (2002): 309–53. https://doi.org/10.1080/0734151022000023785.

8 Joyce Antler, *Jewish Radical Feminism: Voices from the Women's Liberation Movement* (New York: New York University Press, 2018); Jillian M. Hinderliter, 'Muckraking Wonders: Jewish Journalist-Activists of the US Women's Health Movement, 1969–1990.' *American Jewish History* 104, no. 2/3 (April/July 2020): 371–95. https://doi.org/10.1353/ajh.2020.0032.

9 Agnes Arnold-Forster, *The Cancer Problem: Malignancy in Nineteenth-Century Britain* (Oxford: Oxford University Press, 2021), p. 30.

10 David Cantor, 'Cancer: Radical Surgery and the Patient', in *The Palgrave Handbook of the History of Surgery*, edited by Thomas Schlich, pp. 457–77 (London: Palgrave Macmillan, 2018).

11 Maren Klawiter, *The Biopolitics of Breast Cancer: Changing Cultures of Disease and Activism* (Minneapolis, MN: University of Minnesota Press, 2008), Chapters 3–4.

12 Klawiter, *The Biopolitics of Breast Cancer*, Chapters 3–4.

13 On this, see also Barbara Brenner, 'Sister Support: Women Create a Breast Cancer Movement', in *Breast Cancer: Society Shapes an Epidemic*, edited by Ann Kasper and Susan Ferguson, pp. 325–53 (New York: Palgrave Macmillan, 2000).

14 Robert Aronowitz, *Unnatural History: Breast Cancer and American Society* (Cambridge: Cambridge University Press, 2007), p. 265.

15 Bernard Fisher, 'The Evolution of Paradigms for the Management of Breast Cancer. A Personal Perspective.' *Cancer Research* 52, no. 9 (1 May 1992): 2371–83.

16 Fisher, 'The Evolution of Paradigms for the Management of Breast Cancer.'

17 Fisher, 'The Evolution of Paradigms for the Management of Breast Cancer.'

18 D. Cantor, 'Cancer Control and Prevention in the Twentieth Century.' *Bulletin for the History of Medicine* 81, no. 1 (spring 2007): 23. https://doi.org/10.1353/bhm.2007.0001.

19 Richard Doll and Richard Peto, 'The Causes of Cancer. Quantitative Estimates of Avoidable Risks of Cancer in the United States Today.' *Journal of the National Cancer Institute* 66, no. 6 (June 1981): 1226. https://doi.org/10.1093/jnci/66.6.1192. Doll and Peto were discussing the links between diet and cancer aetiology. On this aspect, see Angela Creager, '"EAT. DIE." The Domestication of Carcinogens in the 1980s', in *The Risk on the Table: Food Production, Health and the Environment*, edited by Angela Creager and Jean Paul Gaudillière, pp. 105–37 (New York: Berghahn Books, 2021). On Doll and Peto's report, see Robert Proctor, *Cancer Wars: How Politics Shapes What We Know about Cancer* (New York: Basic Books, 1995).

20 Michael Sporn, Nancy Dunlop, Dianne Newton, and Joseph Smith, 'Prevention of Chemical Carcinogenesis by Vitamin A and Its Synthetic Analogs (Retinoids).' *Federation Proceedings* 35, no. 6 (1976): 1332–8; Michael Sporn, 'Approaches to Prevention of Epithelial Cancer during the Preneoplastic Period.' *Cancer Research* 36, no. 7 (1976): 2699–702.

21 Peter Greenwald, 'Chemoprevention Research at the U.S. National Cancer Institute.' *Military Medicine* 159, no. 7 (July 1994): 505–12.

22 Nancy Tomes, *Remaking the American Patient: How Madison Avenue and Modern Medicine Turned Patients into Consumers* (Chapel Hill, NC: University of North Carolina Press, 2016).

23 Barbara Seaman, *The Doctors' Case against the Pill* (New York: Peter H. Wyden, 1969). On Seaman, see Kelly O'Donnell, 'Our Doctors, Ourselves: Barbara Seaman and Popular Health Feminism in the 1970s.' *Bulletin of the History of Medicine* 93, no. 4 (winter 2019): 550–76. https://doi.org/10.1353/bhm.2019.0072; Kelly O'Donnell, 'The Case against the Doctors: Gender, Authority, and Critical Science Writing in the 1960s.' *Journal of the History of Medicine and Allied Sciences* 75, no. 4 (October 2020): 429–47. https://doi.org/10.1093/jhmas/jraa028; Kelly O'Donnell, 'The Activist Archive: Feminism, Personal-Political Papers, and Recent Women's History.' *Journal of Women's History* 32, no. 4 (winter 2020): 88–109. https://doi.org/10.1353/jowh.2020.0041.

24 Sandra Morgen, *Into Our Own Hands: The Women's Health Movement in the United States, 1969–1990* (New Brunswick, NJ: Rutgers University Press, 2002); Diana Dutton, *Worse than Disease: Pitfalls of Medical Progress* (Cambridge: Cambridge University Press, 1988); Heather Munro Prescott, *The Morning After: A History of Emergency Contraception in the United States* (New Brunswick, NJ: Rutgers University Press, 2011); S. Bell, 'Gendered Medical Science. Producing a Drug for Women.' *Feminist Studies* 21, no. 3 (Autumn 1995): 469–500. https://doi.org/10.2307/3178195. On the early history of diethylstilboestrol in Germany, see Jean-Paul Gaudillière, 'Hormones at Risk: Cancer and the Medical Uses of Industrially Produced Sex Steroids in Germany, 1930–1960', in *The Risks of Medical Innovation: Risk Perception and Assessment in Historical Context*, edited by Thomas Schlich and Ulrich Tröhler, pp. 148–69 (London: Routledge, 2005).

25 Interview with Belle Shayer, 9 June 2017.

26 Flyer, 1990, Box 1, BCA Archive.

27 Flyer, 1990, Box 1, BCA Archive.

28 Anon., 'Network Opposes Tamoxifen Trial.' *Network News*, January–February 1991, p. 3.

29 Anon., 'Network Opposes Tamoxifen Trial.'

30 Anon., 'Network Opposes Tamoxifen Trial.'

31 Adriane Fugh-Berman and Samuel Epstein, 'Tamoxifen: Disease Prevention or Disease Substitution?' *Lancet* 340, no. 8828 (7 November 1992): 1144. https://doi.org/10.1016/0140-6736(92)93161-F.

32 M. Sporn, 'Carcinogenesis and Cancer: Different Perspectives on the Same Disease.' *Cancer Research*, 51, no. 23 (1 December 1991): 6215.

33 Sporn, 'Carcinogenesis and Cancer.'

34 Sporn, 'Carcinogenesis and Cancer.'

35 Anon., 'Tamoxifen Study Group.' *BCA Newsletter*, April 1991, p. 1.

36 Anon., 'Tamoxifen Study Group', p. 1.
37 Anon., 'Tamoxifen Study Group', p. 1.
38 Barbara Brenner, 'Tamoxifen and You – What You Don't Know Can Harm You.' *BCA Newsletter*, February 1996, p. 2; Nancy Evans, 'Tamoxifen Update: Debunking a Wonder Drug.' *BCA Newsletter*, April 1996, p. 1.
39 Nancy Evans, 'National Breast Cancer Awareness Month: Reflections in a Jaundiced Eye.' *BCA Newsletter*, October 1993, p. 6.
40 Evans, 'National Breast Cancer Awareness Month', p. 6.
41 Evans, 'National Breast Cancer Awareness Month', p. 6.
42 Nancy Evans, 'Moving to a Wider Giving.' *BCA Newsletter*, February 1995, p. 2.
43 Anon., 'New BCA Corporate Donation Policy.' *BCA Newsletter*, October/November 1998, p. 5.
44 National Cancer Institute, 'Breast Cancer Prevention Trial Shows Major Benefit, Some Risk. ' 6 April 1998. https://web.archive.org/web/19990221145512/http://rex.nc.nih.gov/massmedia/pressreleases/prevtrial.htm. Accessed 15 July 2022.
45 National Cancer Institute, 'Breast Cancer Prevention Trial Shows Major Benefit, Some Risk.'
46 Barbara Brenner, ' "Prevention Pill" – More Harm than Good?' *BCA Newsletter*, June/July 1998, p. 1.
47 Brenner, 'Prevention Pill', p. 1.
48 Brenner, 'Prevention Pill', p. 1.
49 Brenner, 'Prevention Pill', p. 1.
50 Barbara Brenner, 'Warning: FDA Approves New Label for Tamoxifen.' *BCA Newsletter*, December 1998/January 1999), p. 5.
51 Brenner, 'Warning', p. 5.
52 Brenner, 'Warning', p. 5.
53 Brenner, 'Warning', p. 5.
54 Letter from Barbara Brenner and Jane S. Zones to Richard Klausner, 28 October 1998, Board Meetings 1998, BCA Archive.
55 Daniel Carpenter, *Reputation and Power: Organizational Image and Pharmaceutical Regulation at the FDA* (Princeton, NJ: Princeton University Press, 2010).
56 'AstraZeneca: MYTHS.' March 1999. https://adsspot.me/media/tv-commercials/nolvadex-breast-cancer-drug-myths-21ae9c6d5c1c [last time accessed: 13/1/2022].
57 'AstraZeneca: MYTHS.'
58 'AstraZeneca: MYTHS.'
59 'AstraZeneca: MYTHS.'
60 'AstraZeneca: MYTHS.'

61 Letter from Barbara Brenner to Nora Cody, Judy Norsigian, Cindy Pearson, and Amy Pett, 10 March 1999, Correspondence 1999, BCA Archive.
62 Letter from Barbara Brenner to Nora Cody, Judy Norsigian, Cindy Pearson, and Amy Pett.
63 Letter from Barbara Brenner to Nora Cody, Judy Norsigian, Cindy Pearson, and Amy Pett.
64 Letter from Barbara Brenner to Nora Cody, Judy Norsigian, Cindy Pearson, and Amy Pett.
65 Letter from Barbara Brenner to Nora Cody, Judy Norsigian, Cindy Pearson, and Amy Pett.
66 Letter from Barbara Brenner to Nora Cody, Judy Norsigian, Cindy Pearson, and Amy Pett.
67 Anon., 'Commercial Wins Award.' *Journal of the National Cancer Institute* 92, no. 24 (20 December 2000): 1978.
68 Prevention First Brochure, Campaigns, Prevention First, BCA Archive.
69 Prevention First Brochure.
70 Prevention First Brochure.
71 Prevention First Brochure.
72 Sheryl Ruzek and Julie Becker, 'The Women's Health Movement in the United States: From Grass-Roots Activism to Professional Agendas.' *Journal of the American Medical Women's Association* 54, no. 1 (1999): 4–8.
73 Jean-Paul Gaudillière, 'Conflict of Interest, Interest, Capture and Hegemony in the Diethylstilbestrol Food Crisis', in Creager and Gaudillière (eds), *The Risk on the Table*, pp. 243–73; Jean-Paul Gaudillière, 'Conflits d'intérêts, science industrielle et expertise dans les controversies américaines sur les usages des hormones sexuelles.' *Sciences Sociales et Santé* 38, no. 2 (2020): 21–47. https://doi.org/10.1684/sss.2020.0175.
74 Greene, *Prescribing by Numbers*, p. ix.
75 Sharon Batt, *Health Advocacy, Inc.: How Pharmaceutical Funding Changed the Breast Cancer Movement* (Vancouver: UBC Press, 2017).

Bibliography

Anonymous. 'Commercial Wins Award.' *Journal of the National Cancer Institute* 92, no. 24 (20 December 2000): 1978.
Anonymous. 'Network Opposes Tamoxifen Trial.' *Network News*, January–February 1991.
Anonymous. 'New BCA Corporate Donation Policy.' *BCA Newsletter*, October/November 1998.

Anonymous. 'Tamoxifen Study Group.' *BCA Newsletter*, April 1991.

Antler, Joyce. *Jewish Radical Feminism: Voices from the Women's Liberation Movement.* New York: New York University Press, 2018.

Arnold-Forster, Agnes. *The Cancer Problem: Malignancy in Nineteenth-Century Britain.* Oxford: Oxford University Press, 2021.

Aronowitz, Robert. *Unnatural History: Breast Cancer and American Society.* Cambridge: Cambridge University Press, 2007.

'AstraZeneca: MYTHS.' March 1999. https://adsspot.me/media/tv-comm ercials/nolvadex-breast-cancer-drug-myths-21ae9c6d5c1c. Accessed 13 January 2022.

Batt, Sharon. *Health Advocacy, Inc.: How Pharmaceutical Funding Changed the Breast Cancer Movement.* Vancouver: UBC Press, 2017.

Bell, S. 'Gendered Medical Science. Producing a Drug for Women.' *Feminist Studies* 21, no. 3 (1995): 469–500.

Brenner, Barbara. '"Prevention Pill" – More Harm than Good?' *BCA Newsletter*, June/July 1998.

Brenner, Barbara. 'Sister Support: Women Create a Breast Cancer Movement', in *Breast Cancer: Society Shapes an Epidemic*, edited by Ann Kasper and Susan Ferguson, pp. 325–53. New York: Palgrave Macmillan, 2000.

Brenner, Barbara. 'Tamoxifen and You – What You Don't Know Can Harm You.' *BCA Newsletter*, February 1996.

Brenner, Barbara. 'Warning: FDA Approves New Label for Tamoxifen.' *BCA Newsletter*, December 1998/January 1999.

Cantor, David. 'Cancer Control and Prevention in the Twentieth Century.' *Bulletin for the History of Medicine* 81, no. 1 (2007): 1–38.

Cantor, David. 'Cancer: Radical Surgery and the Patient', in *The Palgrave Handbook of the History of Surgery*, edited by Thomas Schlich, pp. 457–77. London: Palgrave Macmillan, 2018.

Carpenter, Daniel. *Reputation and Power: Organizational Image and Pharmaceutical Regulation at the FDA.* Princeton, NJ: Princeton University Press, 2010.

Creager, Angela, and Jean-Paul Gaudillière. *The Risk on the Table: Food Production, Health and the Environment.* New York: Berghahn Books, 2021.

Doll, Richard, and Peto, Richard. 'The Causes of Cancer. Quantitative Estimates of Avoidable Risks of Cancer in the United States Today.' *Journal of the National Cancer Institute* 66, no. 6 (1981): 1191–308.

Dutton, Diana. *Worse than Disease: Pitfalls of Medical Progress.* Cambridge: Cambridge University Press, 1988.

Evans, Nancy. 'Moving to a Wider Giving.' *BCA Newsletter*, February 1995.

Evans, Nancy. 'National Breast Cancer Awareness Month: Reflections in a Jaundiced Eye.' *BCA Newsletter*, October 1993.

Evans, Nancy. 'Tamoxifen Update: Debunking a Wonder Drug.' *BCA Newsletter*, April 1996.

Fisher, B. 'The Evolution of Paradigms for the Management of Breast Cancer. A Personal Perspective.' *Cancer Research* 52 (1992): 2371–83.

Fosket, J. 'Constructing "High-Risk Women": The Development and Standardization of a Breast Cancer Risk Assessment Tool.' *Science, Technology, & Human Values* 29, no. 3 (2004): 291–313.

Fugh-Berman, A. and Epstein, S. 'Tamoxifen: Disease Prevention or Disease Substitution?' *Lancet* 340, no. 8828 (7 November 1992): 1143–5.

Gaudillière, J.P. 'Conflits d'intérêts, science industrielle et expertise dans les controverses américaines sur les usages des hormones sexuelles.' *Sciences Sociales et Santé* 38, no. 2 (2020): 21–47.

Greene, Jeremy. *Prescribing by Numbers: Drugs and Definition of Disease.* Baltimore, MD: Johns Hopkins University Press, 2007.

Greenwald, P. 'Chemoprevention Research at the U.S. National Cancer Institute.' *Military Medicine* 159, no. 7 (1994): 505–12.

Hinderliter, J.M. 'Muckraking Wonders: Jewish Journalist-Activists of the US Women's Health Movement, 1969–1990.' *American Jewish History* 104, no. 2/3 (2020): 371–95.

Klawiter, Maren. *The Biopolitics of Breast Cancer: Changing Cultures of Disease and Activism.* Minneapolis, MN: University of Minnesota Press, 2008.

Klawiter, Maren. 'Risk, Prevention and the Breast Cancer Continuum: The NCI, the FDA, Health Activism and the Pharmaceutical Industry.' *History and Technology* 18, no. 4 (2002): 309–53.

Löwy, Ilana. *Preventive Strikes: Women, Precancers and Prophylactic Surgery.* Baltimore, MD: Johns Hopkins University Press, 2010.

Maximov, Philipp, Russell McDaniel, and Craig Jordan. *Tamoxifen: Pioneering Medicine in Breast Cancer.* Basel: Springer Basel, 2013.

Morgen, Sandra. *Into Our Own Hands: The Women's Health Movement in the United States, 1969–1990.* New Brunswick, NJ: Rutgers University Press, 2002.

Munro Prescott, Heather. *The Morning After: A History of Emergency Contraception in the United States.* New Brunswick, NJ: Rutgers University Press, 2011.

National Cancer Institute. 'Breast Cancer Prevention Trial Shows Major Benefit, Some Risk.' 6 April 1998. https://web.archive.org/web/199 90221145512/http://rex.nc.nih.gov/massmedia/pressreleases/prevtrial. htm. Accessed 15 July 2022.

O'Donnell, K. 'The Activist Archive: Feminism, Personal-Political Papers, and Recent Women's History.' *Journal of Women's History* 32, no. 4 (2020): 88–109.

O'Donnell, K. 'The Case against the Doctors: Gender, Authority, and Critical Science Writing in the 1960s.' *Journal of the History of Medicine and Allied Sciences* 75, no. 4 (2020): 429–47.

O'Donnell, K. 'Our Doctors, Ourselves: Barbara Seaman and Popular Health Feminism in the 1970s.' *Bulletin of the History of Medicine* 93, no. 4 (2019): 550–76.

Proctor, Robert. *Cancer Wars: How Politics Shapes What We Know about Cancer*. New York: Basic Books, 1995.

Quirke, V. 'Imperial Chemical Industries and Craig Jordan, "the First Tamoxifen Consultant", 1960s–1990s.' *Ambix* 67, no. 3 (2020): 289–307.

Raffensperger, C., and J. Tickner (eds). *Protecting Public Health and the Environment: Implementing the Precautionary Principle*. Washington, DC: Island Press, 1999.

Ruzek, S., and Becker, J. 'The Women's Health Movement in the United States: From Grass-Roots Activism to Professional Agendas.' *Journal of the American Medical Women's Association* 54, no. 1 (1999): 4–8.

Seaman, Barbara. *The Doctors' Case against the Pill*. New York: Peter H. Wyden, 1969.

Schlich, Thomas, and Ulrich Tröhler. *The Risks of Medical Innovation: Risk Perception and Assessment in Historical Context*. London: Routledge, 2005.

Sporn, M. 'Approaches to Prevention of Epithelial Cancer during the Preneoplastic Period.' *Cancer Research* 36, no. 7 (1976): 2699–702.

Sporn, M. 'Carcinogenesis and Cancer: Different Perspectives on the Same Disease.' *Cancer Research* 51, no. 23 (1 December 1991): 6215–18.

Sporn, M., N. Dunlop, D. Newton, and J. Smith. 'Prevention of Chemical Carcinogenesis by Vitamin A and Its Synthetic Analogs (Retinoids).' *Federation Proceedings* 35, no. 6 (1976): 1332–8.

Timmermans, S., and M. Berg. 'The Practice of Medical Technology.' *Sociology of Health and Illness* 25, no. 3 (2003): 97–114.

Tomes, Nancy. *Remaking the American Patient: How Madison Avenue and Modern Medicine Turned Patients into Consumers*. Chapel Hill, NC: University of North Carolina Press, 2016.

8

'Mental health is not fashion': RIP shirts, stigma, and consumerism

Christopher M. Rudeen

In August 2019, *The Washington Post* ran a story on a growing cottage industry around 'Rest in Peace' (RIP) T-shirts, or what are essentially wearable shrines worn to commemorate loved ones lost. These shirts, the author wrote, 'have become a somber material extension of the nation's social epidemics: inner-city gun violence, mass shootings, drug overdoses'.[1] A phenomenon dating back to the late 1980s, such shirts carry many meanings. Todd Smedley – the founder of T-Shirt Kings in Dayton, Ohio – emphasized that the victims 'were still loved', and that his job was to 'help families narrate that story'.[2] Elaborating, Smedley also said that he considered the shirts to be 'some type of therapy' – and he is not alone.[3] Christian Ray, a graffiti artist known as Arson who works at Big City Fashions in Chicago, told a reporter in 2017: 'I'm a graffiti artist, a therapist, a financial advisor, all that'.[4]

RIP shirts, adorned with the pictures or names of those lost too soon, are grave manifestations of the violence of marginalization. These shirts have become part of a growing industry of grief in the wake of national outrage over police brutality and anti-Black racism. But what does it mean for shirts to be 'some type of therapy'? And what, moreover, does it reveal of the deep cultural ambivalences surrounding the use of consumer technologies as a means of healing more generally? By reading RIP shirts through the heuristic tool of family therapy, I will argue in this chapter that the materiality afforded by this technology helps externalize and disperse group trauma, a practice shaped, but not impeded, by larger capitalist regimes. In this sense, those seeking out RIP shirts become unwitting patient consumers, hoping for care while simultaneously contending with the commercial interests that their shirts bring into the fold.

Technology, patient consumers, and medical frameworks

Shirts, for some, may seem disconnected from popular conceptions about technology. 'Technology', however, is notoriously difficult to define, in part because it is simultaneously expansive and narrow in what it encompasses. The simplest definition is perhaps the platonic separation of 'ideas' from 'things', with science representing the former and technology the latter, as suggested by historian Melvin Kranzberg.[5] A longer, yet more complex and traditional, definition is provided by science and technology studies (STS) scholars Wiebe E. Bijker and John Law. As they describe it: 'technologies are not purely technological. Instead ... they are heterogeneous ... artifacts [that] embody trade-offs and compromises. In particular, they embody social, political, psychological, economic, and professional commitments, skills, prejudices, possibilities, and constraints'.[6] This definition indicates that technology is not just something material but also a vessel for ideas, a marker of relationships, and an instrument for achieving certain ends. In this vein, shirts are assuredly a form of technology: they are multifaceted objects that help individuals meet their goals.

In a book on patient consumers, it might also be unexpected to include people who purchase shirts. They are not patients in the direct sense of being individuals who are sick and subject to medical diagnosis and treatment. Categorizing them as patients and using psychiatric terminology to analyse the social problems that they are part of, moreover, are in some ways acts of medicalization. This is potentially even more problematic because these individuals tend to be multiply disenfranchised, and one must be wary of pathologizing individuals who are not sick but suffering from social injustice.[7] However, there is an analytical and political benefit in engaging in such work. It makes sense to consider communities harmed by 'social epidemics' such as gun violence and drug overdoses, not to mention the negative psychological impacts of structural racism, as being composed of individuals who are suffering and seeking care; in particular, it allows us to expand our conception of who gets access to labels such as 'patient'.[8] Given the many inequalities present in medical encounters, limiting one's focus to solely the spaces of hospitals, doctors' offices, and pharmacies too often excludes certain groups of people.[9]

These frameworks aside, 'RIP' or 'memorial' shirts have received scant attention from academics. Several young scholars, most notably Katie Kavanagh O'Neill and Robin Brooks, have focused on their use in practices of mourning. Kavanagh O'Neill – whose dissertation is a people's history of violence in Baltimore during the War on Drugs – sees such shirts as memorializing those murdered in the city.[10] Robin Brooks, meanwhile, reads them as 'visual life writing' that tell stories both collective – about the Movement for Black Lives, for instance – and individual.[11]

To further contextualize the many meanings of these cultural objects, this chapter draws on work in family therapy, psychology, studies on stigma, and the history of medicine. Breast cancer and HIV/AIDS are two diseases that have been heavily stigmatized, and secondary literature on the ways in which different communities have sought to cope with that stigma gives insight into the case at hand. Research into cancer, especially on breast cancer, has shown how the disease's identity has changed from a private experience to a collective risk and how different groups of sufferers have fought to make their pain legible.[12] The AIDS quilt is another example of health activism, in this case undertaken in response to that pandemic in the United States. Individual sufferers and their communities worked to cultivate expertise, which was then harnessed for both medical and societal advances in treatment of the disease and those communities themselves.[13] As will be discussed later, these are also examples in which wearable technologies became outward symbols of the cause, as well as contested objects of support.

Using newspaper reports, vendors' websites, and secondary literature on breast cancer and HIV/AIDS activism, this chapter thus teases apart the therapeutic potential of RIP shirts. Divided into three main parts, the chapter's first section employs the insights of structural family therapy, especially the concept of the 'identified patient', to show how the story of one member of a group is used to understand broader 'social epidemics'. The second looks at RIP shirts as diffusible stigma, a way of externalizing and reframing stigmatized positionalities. Namely, since therapeutic action consists of shocking the system to change its arrangement, these shirts attempt to do so by extending the community of mourners in the hope of changing how victims of drug overdoses or police brutality are seen. The final section turns to the ambivalences of the shirts and the role they play in wider consumer culture, highlighting their limitations and unique advantages. When consumption is local, makers can help in crafting new narratives and community support, whereas when the practice is purely commercial, the message, and money, can be lost.

Finally, this inquiry deals with only one use of RIP shirts: those recording 'spectacular' deaths related to the 'social epidemics' described above.[14] Images of the deaths of people such as Stanley 'Tookie' Williams, Trayvon Martin, and Eric Garner have circulated widely, becoming, in the words of art historian Nicole R. Fleetwood, 'troubling' images.[15] In general, RIP shirts are not linked to such public figures and are instead profoundly personal objects that do not travel very widely. Other scholars, most notably Rikki Byrd, focus on the more private memorialization the shirts enact in those cases. This chapter focuses less on the shirts' images or these practices than on their role as a technology doing certain kinds of work: linking, through consumption, a group of individuals in search of healing.

'The individual patient is not the patient': family therapy and structures of healing

Family therapy, like Tupperware and the atomic clock, was a postwar invention. Thinkers in a range of disciplines turned their attention to the family, where, it was hoped, the ills of broader society could be located and treated. Historian Deborah Weinstein lists the three largest influences on this nascent field as cybernetics, systems theory, and the 'culture' concept being developed in the social sciences.[16] This is especially clear in structural family therapy, a major strand of the field associated with practitioner Salvador Minuchin. By applying some of the tools of this therapeutic modality – the concept of the 'identified patient', an understanding of behaviour as a 'shared responsibility', and the goal of structural change – to RIP shirts, we learn how the garments can materialize social ills and point to new ways of relating that begin to address long-standing trauma. In short, structural family therapy conceptualizes pathology as inherent within groups of people, not individuals. The figure immortalized on a shirt thereby highlights societal ills.

This process also invites the makers of RIP shirts into the family group. While I am using family therapy as a heuristic device, I want to emphasize that it is still a logical step to apply the theory to the case at hand, especially when using a heterogeneous understanding of the word 'family'. The term carries different meanings depending on the setting.[17] For Minuchin, the family was defined as 'a special kind of system, with structure, patterns, and properties that organize stability and change'. The family was also 'a small human society, whose members have face-to-face contact, emotional ties, and a shared history'.[18] 'The family' is thereby not necessarily synonymous with the nuclear family, especially for specific cultures.[19]

The goals of structural family therapy are simple in theory. The first is 'challenging the IP's [identified patient's] ownership of the symptom', followed by 'making the family members responsible for the maintenance of the symptom'.[20] One of the core concepts of family therapy is the 'identified' (or 'individual') patient. Unlike traditional psychotherapy, family therapists argue that, in the words of Minuchin, '[t]he individual patient is not the patient. The patient is the relational patterns that are created by belonging to a subsystem'.[21] The 'identified patient' thus expresses the pathology of the family and its relations. The final aim of the therapy is changing the family's organization; when this is achieved, individual experience shifts as well.[22]

The figure emblazoned on the RIP shirt is akin to the identified patient. The acts of printing and wearing transform the individual from victim to symptom of the violent relational patterns created by belonging to a minoritized group in a society shaped by structural racism. The problem is in the relations *among* people and not *in* any one individual, highlighting the broader structural system in need of 'cure'. Those who wear the shirts become 'patients', too, suffering from both the loss of loved ones and the ongoing effects of the system that killed them.

For months after George Floyd's murder in May of 2020, the site of his death became an impromptu memorial. Many flocked to the area, and merchandise featured heavily. One vendor, Clifford Dodd, sold clothes adorned with the inscription 'Black Lives Matter' and 'We Still Can't Breathe'. 'Everybody has their own stories of being caught by the cops, being degraded', Dodd told reporters, 'and everyone came here and they were laying their pain down somewhere'. Krystel Smith, who tended to the flowers and lit candles for the dead, wore one such shirt, saying: 'I just get sad and emotional, and I get to thinking about everything like, damn, this could have been my brother, this could have been my dad, this could have been my boyfriend. *He could've been anybody*'.[23] The site provides an occasion to recognize shared pain resulting from larger societal structures of anti-Blackness, the violence that arises from, in the words of scholar kihana miraya ross, 'the inability to recognize black humanity'.[24] This is not to lose sight of the individual, but to acknowledge that the problem is not located in the individual. The problem is a consequence of belonging in the American family as it now stands.

Structural family therapy also has implications for understanding behaviour. The theory describes behaviour as 'a shared responsibility, arising from patterns that trigger and maintain the actions of each individual' within a family.[25] Behaviour is never one-sided; instead, as Minuchin put it, 'the process [of interaction] is *circular* and the behavior is *complementary*'.[26] Minuchin gave the following example from work with families in the welfare system:

> We can say with equal validity that, when Tamika is defiant, her mother yells, Tamika cries, and her mother hits her – or, that mother yells at her daughter, Tamika cries, her mother hits her, and Tamika becomes defiant. Their interaction is patterned, and we cannot explain the behavior of one without including the other.[27]

As behaviour is patterned in interactions, so too is race. In a review of Karen E. Fields and Barbara J. Fields' *Racecraft*, sociologist Ruha Benjamin explained the concept's mystifying power to confound race and racism with a similar example, worth quoting at length:

> In the wake of the murder of Michael Brown in Ferguson, Missouri, one might describe events as 'an unarmed teenager was shot because he was black'. Racecraft converts power into difference insofar as the young man's race, his *being black*, is given agency – an ontology – thereby veiling the work of multiple forms of racism that led a law enforcement official to shoot this young man to death. The two Fields would remind us that Brown's blackness did not pull the trigger, as the original formulation mistakenly asserts. Rather, deeply

entrenched forms of ideological paranoia and institutional pathology, which routinely conflate blackness with criminality, ensured that officer Darren Wilson would treat Brown as a threat. In a social context characterized by racecraft, causal relationships are typically stated incorrectly. Brown was not shot because he was black. *He is black because he was shot.*[28]

While Benjamin refuses to accept reciprocity between Brown and Wilson, both examples indicate how behaviour cannot be explained by the actions of one individual. It was 'deeply entrenched forms of ideological paranoia and institutional pathology' that led to the murder. Michael Brown was the identified patient, but the 'paranoia' and 'pathology' lay in society.

Treatment by structural family therapy proceeds via 'therapeutic interventions that confront and challenge a family in the attempt to force a therapeutic change'.[29] Take Minuchin's work with the B family, whose eldest son Stephen presented with anorexia nervosa. The therapists gave the household a series of tasks. The tasks induced a crisis that demonstrated 'the extent to which a problem that had been conceptualized as Stephen's was built into the structure of the family'. Subsequent interventions focused instead on the family's negotiation of conflict.[30]

I contend that RIP shirts, in creating community through sartorial attachment, participate in this type of structural change. The memorial shirt emphasizes the shared nature of grief, inviting others who did not know the deceased personally to participate in mourning. Kyle Ng, one of the founders of Brain Dead, commented on the label's version of an RIP shirt – which read: 'If you love black culture, protect black lives' – by saying: 'The world is our family and we've got to help them out.'[31] The shirt, especially as a form of streetwear, acts as a material form of community.[32]

Objects are an underutilized but vastly productive means by which to understand the self and its relation to others. In 1988, business professor Russell W. Belk outlined what he called the 'extended self'. Belk argued that consumer behaviour cannot be understood without understanding the relationship between people and goods, namely the idea that 'knowingly or unknowingly, intentionally or unintentionally, we regard our possessions as parts of ourselves'.[33] Moreover, Belk argued that these possessions 'give us *a personal archive or museum* that allows us to reflect on our histories and how we have changed'.[34] In particular, clothes can be considered a special case of what Ann Cvetkovich has termed an 'archive of feelings', a group of 'cultural texts [explored] as repositories of feelings and emotions, which are encoded not only in the content of the texts themselves but in the practices that surround their production and reception'.[35] Items such as the RIP shirts participate in imagining new ways of healing from trauma and caring for others.

Here, the example of the AIDS quilt is instructive. Conceived in 1985, the NAMES Project Foundation's AIDS Memorial Quilt was first displayed in Washington, DC, in 1987. Its initial iteration included 1,920 panels, each representing a person who had lost their life to the disease.[36] Its demonstration brought the domestic space where quilts are usually found into the public sphere. In his study of the project, Peter Hawkins noted that Cleve Jones alighted on the quilt as a 'domestic equivalent for the sign of national unity', a 'metaphor of *e pluribus unum* that was (to recall his words) cozy, humane, and warm'.[37] Displayed on the National Mall, the quilt offered some 'a way to suffer intimate losses in the most public space in America, to leave behind ghetto and closet, to bring mourning from the margin to the center'.[38] Steven James Gambardella went further, arguing the quilt 'transformed the mainstream', changing public perceptions of the AIDS-affected community and redefining what the 'public' actually was.[39] This piece, as part of an 'archive of feelings', allowed for the creation of new 'public cultures' around group trauma. As Cvetkovich wrote, 'The formation of a public culture around trauma has been especially visible in the queer response to the AIDS crisis. Queer activism insisted on militancy over mourning, but also remade mourning in the form of new kinds of public funerals and queer intimacies.'[40]

The haunting power of past and current racial trauma is made material in the RIP shirt, which, following Cvetkovich, 'enables new practices and publics'.[41] RIP shirts show the radical relationality present in social trauma, locating structural pathology while suggesting new ways of belonging and healing.

'A first impression of our fellow-creatures':
stigma and relationality

I have argued that objects such as T-shirts can function as material representations of internal and interpersonal processes. This section expands on this idea to look more deeply at a possible mechanism by which clothing can act in group healing. As a boundary object between self and other, clothing has the power to externalize and concretize internal processes. In this way, it is a technology that can diffuse stigma, allowing groups to defuse harmful associations by adding more and more members to the cause. There is a price to this endeavour, however, as it binds healing to the consumer realm and the sometimes-hidden forces contained therein.

Susan Leigh Star and James R. Griesemer developed the concept of the 'boundary object' to describe 'the *flow* of objects and concepts through the *network* of participating allies and social worlds' without narrowing the focus to any one point of view.[42] These social worlds are held together in part through boundary objects, which are 'both plastic enough to adapt to local needs and the constraints of the several parties employing them, yet robust enough to maintain a common identity across sites'.[43] Clothing is a boundary object *sui generis*. Many modalities of psychology and psychiatry argue that objects are crucial to understanding the self and its relation to others.[44] As British psychologist and psychoanalyst J.C. Flügel explained it in a book titled *The Psychology of Clothes* in 1930, clothes, 'though seemingly mere extraneous appendages, have entered into the very core of our existence as social beings'. Clothes are central both to social intercourse, providing 'a first impression of our fellow-creatures as we meet them', and to 'the core of our existence' as individuals presenting such impressions to others.[45]

If objects can be part of the self, they can also be manipulated to aid in treatment – including the reduction of grief and stigma. Memorial shirts and related items of clothing, I argue, can in fact function as mobile stigmata that externalize and redefine group boundaries, promoting healing by changing the relationship between visible attributes and perceived desirability.[46] Historian Samantha King, in her book on the cultural transformation of breast cancer in the late-twentieth-century United States, traces a similar path in the work of activists redefining the experience of the disease. Breast cancer, King wrote, 'has been reconfigured from a stigmatized disease and individual tragedy best dealt with privately and in isolation, to a neglected epidemic worthy of public debate and political organizing, to an enriching and affirming experience during which women with breast cancer are rarely "patients" and mostly "survivors"'. As she points out, the figure in the final stage occasions an outpouring of 'American generosity', one which becomes part of the treatment.[47] The wide variety of items related to the cause – be they pink ribbons or bracelets or shirts – are, not unlike RIP shirts, marketed as helping 'raise awareness' and fund research into a cure.[48]

The act of 'raising awareness' is thus mediated through the creation of community. As we saw above, the lens of family therapy allows us to appreciate how memorial shirts engender self-reflection about forms of relatedness. This is also apparent in the case of breast cancer. King uses the term 'breast cancer culture' to describe 'those forms of breast-cancer-related social identification, affiliation, and organization that exist outside of the confines of social movement organizations or political action and change narrowly defined'.[49] This focus on culture allows King 'to highlight the distinct set of signs and symbols – *style*, even – now associated with' breast cancer.[50] These 'styles' – note the word choice – help to create forms of collective experience. This is most evident in the masses of pink clothing present at the campaign's signature event, a walk now known as 'More than Pink'. The imagery of the Seattle walk in 2017 is a case in point. A photograph on the chapter's website features walkers in all shades of pink under a banner proclaiming: 'Be More than Pink'.[51] United through reference to a single cause and visually in a particular style, this community emphasizes how 'pink' is a way of *being* and, more importantly, of *being together*. The process of destigmatization occurs partly as an outcome of this spread. The distribution of pink shirts shifts focus from one to many, hiding one's markings in the process of creating new ones – it is not clear who in the photograph is a survivor or a supporter, but both are clearly identifiable by a manufactured link to the disease. The externalization and diffusion of the stigma thus allow for its removal, at least in theory.

RIP shirts, I argue, act through similar channels to call attention to the neglected 'social epidemic' of anti-Black violence. Put
simply, the proliferation of shirts extends the community who
mourns such losses. Clothing is a particularly powerful means
by which to do this work for many reasons, not only because
clothing is seen as an extension of self but also because the sale
of items such as shirts produces real material gains for anti-racist
causes.

RIP shirts, in particular, build on an established market for political shirts that dates to the post-war period. The affordability and
ubiquity of shirts, in addition to their simplicity, made them ideal
canvases for communications of all stripes, an explicit rendering
of Flügel's theories.[52] A 1973 article in the *New York Times*, for
instance, featured a collage of shirts under the headline 'The T-Shirt
Has Become the Medium for a Message'. The opening of the article
quotes artist Anne Drew Fox saying that the shirt is a 'nonfash-
ion': 'All it does is cover you. What's interesting about it is the
message it communicates'.[53] Current advertisements for companies
such as Bonfire that specialize in selling shirts for charitable causes
emphasize the power held by shirts to do this work, as well as the
ease of doing so.[54]

This is clearly communicated by those who make RIP shirts. The
first thing a potential customer sees on the website for the Dayton
small business T-Shirt Kings is consolation: 'Losing someone special in your life is tragic', they note, 'and we are extremely sorry
to know this. But this is how life works. You can only remember
the good times you had together and life goes on. Though, you
can keep them alive in your memories and also through memorial
t-shirts'.[55] The shirts 'keep them alive' via relationality: 'Whether
you are planning on a get together in the memory of the deceased or
just want to make them a part of your life when they are no longer
here, memorial t-shirts are a good way to do so.'[56] Through all-day
wear, as a makeshift uniform at memorial gatherings, and by moving through spaces in which other people will see them, the shared
garment literally brings people together to bring about healing. The
shirts are made to be seen by others, who in the act of looking participate in the act of remembrance. The sellers make clear that even
in a drawer, the shirt will make the loved one 'a part of your life
when they are no longer here'.[57]

RIP shirts' emphasis on visible community thereby helps cultivate resilience and, in this way, provide a wide-reaching therapeutic technology helping to overcome traumas experienced by entire communities and groups. Equally, the shirts also act in a manner similar to other therapeutic modalities tailored to African Americans. Therapists such as Ronald Fudge and Beverly Greene, for instance, note that culturally literate psychotherapy must also consider history and society, with both emphasizing the link between positive ethnic identity and resilience.[58] Fudge's developmental model of the positive ethnic persona calls upon William E. Cross Jr's concept of 'nigrescence', which describes the process by which one develops a positive self-identity in opposition to prevailing negative stereotypes.[59] Fudge combines this developmental progression with behavioural therapy to provide patients with a reservoir of strengths stemming from their heritage.[60] This is akin to the process above, where the memorial shirt is presented as a way to focus on happy memories and community to help life go on.

Yet the sale of the shirts is also undeniably big business, providing money both for vendors and broader anti-racist causes that may not necessarily or always be aligned. Reports on the shops in New Orleans indicate that RIP shirts are the backbone of many small businesses. Lawrence Elzy, the owner of Exclusive Tees, told a filmmaker: 'You can survive without doing Rest In Peace shirts, but your business will never grow.'[61] In 1998, another New Orleans shop owner estimated that 90 per cent of his business came from 'dead man shirts'.[62] *The Wall Street Journal* reported that brands were 'turning standout designs into five- and even six-digit fundraisers for causes' in the summer of 2020. Smaller brands, instead of donating to causes like larger labels, 'leveraged their modest clout by selling message tees'.[63]

Yet accompanying the shift from stigmatized condition to cause célèbre is also the rise of larger corporate interests. As King noted, commodities surrounding the breast cancer movement such as the Race for the Cure (as it used to be known) and its shirts 'transform purchasers into certain kinds of people living certain kinds of lives', producing neoliberal 'consumer-citizens'.[64] The same process is happening in the world of memorial tees. Small shops are not the only ones selling RIP shirts, which introduces new questions about the commodification of pain and healing.

'Mental health is not fashion': consumerism and commodification

Gucci opened its 2019 Fashion Week show in Milan with a statement. Models, dressed in white jumpsuits resembling straitjackets, were lined up on a conveyer belt, everything around them stark white. According to designer Alessandro Michele, this 'most extreme version of a uniform dictated by society' was supposed to represent a blank slate, how 'through fashion, power is exercised over life, to eliminate self-expression'.[65] The bold colours and patterns that followed were meant to emphasize the opposite: as the designer told *The New York Times*, 'I wanted to show how society today can have the ability to confine individuality and that Gucci can be the antidote'.[66] One model, however, had a statement of their own. Ayesha Tan-Jones wrote 'mental health is not fashion' on their hands and held them up as a form of silent protest. Tan-Jones wrote later that 'Presenting these struggles as props for selling clothes in today's capitalist climate is vulgar, unimaginative and offensive.'[67]

'Today's capitalist climate' has loomed large in this chapter. Tan-Jones' protest dramatically illustrates, however, the complex effects of consumer capitalism. Yes, the addition of corporate interests has led to conflict, as fashion and the broader society it represents can erase the individual. However, clothing, as an object simultaneously public and private, *can* also offer an antidote. Clothes are a language, allowing, not least, for expression of our innermost desires and feelings.[68] As the *New York Times* article on T-shirts from 1973 stated, 'the T-shirt, then, is a medium for a message and fads in messages come and go'.[69] Here, again, clothing is a boundary object: between public and private, individual and society, self and market.

Historian Nancy Tomes argued in her 2016 monograph on the patient consumer that the intertwining of medicine and business is not new. The seemingly modern idea of 'critical consumerism', a movement to 'protect and inform consumers', in fact dates to Progressive era worries over a proliferation of new medical technologies and claims. While keeping faith in the power of the well-informed patient, Tomes argues that this critical consumerism is plagued by 'the difficulty of separating it from its uncritical counterparts', from other meanings of consumerism. Citing the *American Heritage Dictionary*, Tomes identifies two other definitions: 'The theory that a progressively greater consumption of goods is economically beneficial' and 'Attachment to materialistic values or possessions'.[70]

Memorial shirts, I argue, are subject to the same inextricability from these different strands of consumer logic. The example sentence for the final definition, '*deplored the rampant consumerism of contemporary society*',[71] indicates the view that such attachment is 'deplorable'. However, some theorists argue that attachment is necessary to human functioning and that attachment to possessions is an important step in developing 'healthy' relationships.[72] Clothing and objects present opportunities for mobile forms of self-expression, and in so doing both stabilize our sense of selves and, as political theorist Bonnie Honig argues, become 'public things' that are essential to the work of democracy.[73] 'At their best', Honig wrote, 'public things gather people together, materially and symbolically, and in relation to them diverse peoples may come to see and experience themselves – even if just momentarily – as a common in relation to a commons, a collected if not a collective'.[74]

Memorial shirts are an example of a 'public thing' creating communities through memorialization. A newspaper story on RIP shirts from 2017 opened with a profile of Quentin Harris, whose brother Julian was shot and killed seven years prior. While Quentin was in prison and unable to mourn his brother at the time, he now holds annual celebrations, each with a new memorial shirt, as part of 'the unending process of grieving'. Quentin noted that his brother's children might only remember their father through these customized shirts. 'It feels like I'm giving him a second life,' Harris said. 'Like he'll never really be gone as long as I can help it.'[75] The shirts are meant to be looked at, held, kept. As one designer noted, he felt the 'biggest way [he] could reach people was to make merchandise that's obtainable … someone could see someone else wearing it and it could go on and on'.[76] The shared and material attributes of these shirts thereby create very real relationships.

At the same time, the 'public' or expanded family system these things create necessarily includes businesses selling the shirts. This addition is an uneasy one when those behind the shirts are not readily apparent. Nearly every newspaper article published about RIP shirts includes some form of the question, in the words of one head-line, 'Is It OK to Sell Breonna Taylor T-Shirts?' When do attempts to raise awareness cross the line and become commodification?[77] While it is often the case that local businesses escape this question, these shirts always hold this contradiction between cooperation and co-optation.

The summer after Elijah McClain was killed in Aurora, Colorado, in 2019, a local artist created a widely shared illustration of McClain emphasizing his 'innocence and gentleness', designed to be sold as prints or on shirts to support McClain's family.[78] While this artist reached out to Elijah's mother before creating the image, Sheneen McClain had mixed feelings about people using her son's likeness. After thanking those 'still supporting and fighting for Elijah's jus-tice' on the GoFundMe page on 26 February 2021, Sheneen turned her attention to other members of the community: 'To the individu-als that are still trying to profit from my son's murder, may you also be judged by Heavenly Father's Divine Judgements. Stop asking me about giving my approval for films, documentaries, books, or any-thing else with my son's name attached to it!'[79] The question of who was welcome to mourn alongside Elijah's family was thereby tied to questions of profit.

In a similar vein, *Los Angeles Times* editor Justin Ray wrote about his experience shopping for a face mask featuring Breonna Taylor's name during the COVID-19 pandemic. 'As I searched for the right mask to purchase to make my own statement', Ray wrote, 'I quickly made a realization: I had no idea where the money would be going'. Ray was discouraged to find 'a whole e-commerce ecosystem … monetizing the deaths of Black people', one that 'seems focused on making sales, not making a difference'. Ray recalled thinking: 'What the hell am I doing? This was a real person who suffered a brutal death. Why am I thinking about fashion?' In the end, Ray decided to focus his efforts elsewhere, through his work, relationships, and vote.[80] Similar articles have been written about 'Cashing in on George Floyd',[81] and news outlets reported on a fight over the sale of memorial shirts for Michael Brown, one that included Brown's family – another private event turned media spectacle.[82]

The example of the AIDS Memorial Quilt illustrates that it is in part the practice of handiwork that invests meaning into an object. There is a long but neglected history of patient crafting, from asylum art therapy to the various handmade hearts in the collection of the Texas Heart Institute.[83] In this vein, Ray suggested that 'one potential workaround is to make your own T-shirt or mask'.[84] A newspaper article on a Bronx shirt business that saw increased interest during the summer of 2020 noted that 'It's possible to order and design shirts online from a variety of vendors, but calling or entering the Greenes' store lends a more intimate feel to the process.' As proprietor Karen Greene put it, 'when they see the shirts, it helps the mourning, it's kind of a way of expressing their loss'.[85] This image, of entering a local shop and having a hand in the shirt's production, differs wildly from the experience Ray had shopping online. It illustrates how handiwork's deep connections to community aid the healing process.

The status of RIP shirts as consumer goods thus ultimately raises questions of ownership: who owns these shirts? Who owns this grief? Here we see the constraints of expanding the family system through material means. This consumerism is ultimately of limited use because the economic system is structured by the same forms of systemic inequality that often occasion the shirts in the first place. As the authors of a study on the relationship between economic inequality and firearm homicide wrote, 'Violence, as a major health problem in the U.S. and a sensitive indicator of social relations', has been found to have 'a strong relationship' with income inequality. Such studies, they continued, emphasize 'the importance of policies that attempt to *shift* the underlying societal forces giving rise to firearm violence at the population level', including reducing the income gap.[86] Therefore, while shirts allow one to make a public statement, they also tap into broader neoliberal trends that privilege money over movements. For example, journalist Justin Ray also brought up the case of Amazon, which sells some of these products. Amazon also owns Whole Foods, which was being sued by workers prohibited from wearing Black Lives Matter masks to work.[87] For these 'public things' to imagine new publics, they must confront this system.

Conclusion: 'just a shirt'

RIP shirts are productively unstable objects. Some of the designers refer to their work as 'some type of therapy', and this chapter has used structural family therapy to argue that what is 'treated' by the shirts is the disordered relational patterns between people. This also suggests a possible point of intervention. Memorial shirts invite more people to mourn those lost. This keeps individuals alive, while at the same time highlighting their inclusion in a broader system of suffering that requires intervention. However, these shirts run the risk of commodifying death itself and reinscribing larger inequalities.

One of the advantages of the family therapy lens is the avoidance of (re-)pathologizing minoritized groups. Following the motto that 'the individual patient is not the patient', dysfunction is found in the system and not the individual. As Elijah McClain's mother said, 'We have so much catching up to do as a Community on Earth, so get close to the trustworthy and dependable, leave procrastination in the paragraphs but change your behaviors of existence. To truly have our best change possible, we need to stop inhumane laws and lawmakers.'[88] This structural lens, however, has its drawbacks. As Minuchin wrote, one such difficulty is a loss of control 'consequent upon the broadening of the therapist's field. Because his areas of intervention are spread, his power in each area is diminished'.[89]

As this and other chapters in this volume make clear, suffering is a complex medical and institutional problem. Especially for those distrustful of the medical establishment, consumer medical goods present a meaningful alternative to care received within the walls of the clinic. This is especially so for mental health care, a field with pervasive disparities in access and usage.[90] However, expanding the doctor–patient interaction opens space for other interests – to quote medical historian David J. Rothman, more 'strangers at the bedside'.[91] These memorial shirts are another example of the potential dangers of commercial interests in matters of health. In particular, they illustrate how damage arises from separating context from care. One can purchase a shirt with a picture of Elijah McClain – the same one his father LaWayne Mosley wore at a protest on 1 October 2019 – on Amazon.[92] There is no information on McClain, nor the seller. While Elijah's name is clear, the shirt's description also includes the words 'fashionable', 'trending', and 'funny'.[93] While much work remains to be done on the RIP shirt, one thing remains abundantly certain: in the words of William 'Surf' Ryals, the founder of TB Customs in Chicago, 'it's more than just a shirt'.[94]

Notes

1 Rachel Siegel, 'Where Gun Violence Abounds, Honoring Loved Ones with "Rest in Peace" Shirts.' *Washington Post*, 10 August 2019. https://web.archive.org/web/20200312223546/www.washington post.com/business/2019/08/10/mass-shootings-everyday-grief-honor ing-loved-ones-with-rest-peace-shirts/.
2 Siegel, 'Where Gun Violence Abounds.'
3 Siegel, 'Where Gun Violence Abounds.'
4 Jasmine Sanders, 'Memorial T-Shirts Create a Little Justice, a Tiny Peace.' *New York Times*, 14 November 2017. www.nytimes.com/2017/11/14/style/memorial-t-shirts.html.
5 Melvin Kranzberg, 'At the Start.' *Technology and Culture* 1, no. 1 (1959): 4.
6 Wiebe E. Bijker and John Law , 'Introduction', in *Shaping Technology/ Building Society: Studies in Sociotechnical Change*, edited by Wiebe E. Bijker and John Law, pp. 7–8 (Cambridge: MIT Press, 1992).
7 This is especially important given the long history of such medicalization by the field of psychiatry itself. On the harmful effects of psychiatric medicalization, see, among others, Ronald Bayer, *Homosexuality and American Psychiatry: The Politics of Diagnosis* (New York: Basic Books, 1981); Jonathan Michel Metzl, *Prozac on the Couch: Prescribing Gender in the Era of Wonder Drugs* (Durham, NC: Duke University Press, 2003); Jonathan Michel Metzl, *The Protest Psychosis: How Schizophrenia Became a Black Disease* (Boston, MA: Beacon Press, 2010); Martin Summers, *Madness in the City of Magnificent Intentions: A History of Race and Mental Illness in the Nation's Capital* (New York: Oxford University Press, 2019).
8 Zinzi D. Bailey, Nancy Krieger, Madina Agénor, Jasmine Graves, Natalia Lionos, and Mary Bassett, 'Structural Racism and Health Inequities in the USA: Evidence and Interventions.' *Lancet* 389 (2017): 1453–63.
9 Jonathan Michel Metzl and Helena Hansen, 'Structural Competency: Theorizing a New Medical Engagement with Stigma and Inequality.' *Social Science & Medicine* 103 (2014): 126–33.
10 Katie Kavanagh O'Neill, 'Mobtown Memories: Towards a People's History of Violence in Baltimore' (PhD dissertation, University of Pittsburg, 2017).
11 Robin Brooks, 'R.I.P. Shirts or Shirts of the Movement: Reading the Death Paraphernalia of Black Lives.' *Biography* 41, no. 4 (Autumn 2018): 807–30.

12 My thinking about cancer is informed by Robert A. Aronowitz, *Unnatural History: Breast Cancer and American Society* (New York: Cambridge University Press, 2007); Ilana Löwy, *Preventive Strikes: Women, Precancer, and Prophylactic Surgery* (Baltimore, MD: Johns Hopkins University Press, 2010); Keith Wailoo, *How Cancer Crossed the Color Line* (New York: Oxford University Press, 2012).

13 Steven Epstein, *Impure Science: AIDS, Activism, and the Politics of Knowledge* (Berkeley, CA: University of California Press, 1996).

14 Brooks, 'R.I.P. Shirts', 814–21.

15 Nicole R. Fleetwood, *Troubling Vision: Performance, Visuality, and Blackness* (Chicago, IL: University of Chicago Press, 2011), pp. 8–9.

16 Deborah Weinstein, *The Pathological Family: Postwar America and the Rise of Family Therapy* (Ithaca, NY: Cornell University Press, 2013).

17 Weinstein, *The Pathological Family*, pp. 6–8.

18 Patricia Minuchin, Jorge Colapinto, and Salvador Minuchin, *Working with Families of the Poor*, 2nd edition (New York: Guilford Press, 1998), p. 15.

19 Therapist Nancy Boyd-Franklin noted that many scholars consider the boundaries of Black families to be 'more permeable, more open to outside influence'. Nancy Boyd-Franklin, *Black Families in Therapy: Understanding the African American Experience* (New York: Guilford Press, 2003), p. 305.

20 Salvador Minuchin, 'Deconstructing Minuchin.' *Journal of Systemic Therapies* 36, no. 4 (2017): 96.

21 Minuchin, 'Deconstructing Minuchin', 95–6.

22 Salvador Minuchin, *Families and Family Therapy* (Cambridge, MA: Harvard University Press, 1974), p. 2.

23 Daniella Silva and Ed Ou, 'Three Months after George Floyd's Killing, Memorial Remains "Sacred Place" for Racial Justice.' NBC News, 25 August 2020. www.nbcnews.com/news/us-news/three-months-after-george-floyd-s-killing-memorial-remains-sacred-n1238084. Emphasis added.

24 kihana miraya ross, 'Call It What It Is: Anti-Blackness.' *New York Times* , 4 June 2020. www.nytimes.com/2020/06/04/opinion/george-floyd-anti-blackness.html.

25 Minuchin, Colapinto, and Minuchin, *Working with Families of the Poor*, p. 19.

26 Minuchin, Colapinto, and Minuchin, *Working with Families of the Poor*, p. 19, emphasis in original.

27 Minuchin, Colapinto, and Minuchin, *Working with Families of the Poor*, pp. 19–20.
28 Ruha Benjamin, 'Conjuring Difference, Concealing Inequality: A Brief Tour of Racecraft.' *Theory and Society* 43, no. 6 (November 2014): 684. Emphasis in original.
29 Minuchin, *Families and Family Therapy*, p. 138.
30 Salvador Minuchin, 'The Use of an Ecological Framework in the Treatment of a Child', in *The Child in His Family*, Vol. 1, edited by E. James Anthony and Cyrille Koupernik, pp. 41–57 (New York: Wiley, 1970).
31 Jacob Gallagher, 'These T-Shirts Are Raising Serious Funds for Antiracism Causes.' *Wall Street Journal*, 16 June 2020. www.wsj.com/articles/these-t-shirts-are-raising-serious-funds-for-black-aligned-causes-11592335526.
32 See Bobby Hundreds, *This Is Not a T-Shirt: A Brand, A Culture, A Community – A Life in Streetwear* (New York: Farrar, Straus and Giroux, 2019).
33 Russell W. Belk, 'Possessions and the Extended Self.' *Journal of Consumer Research* 15, no. 2 (September 1988): 139.
34 Belk, 'Possessions and the Extended Self', 159. Emphasis added.
35 Ann Cvetkovich, *An Archive of Feelings: Trauma, Sexuality, and Lesbian Public Cultures* (Durham, NC: Duke University Press, 2003), p. 7.
36 National AIDS Memorial, 'The History of the Quilt .' www.aidsmemorial.org/quilt-history.
37 Peter S. Hawkins, 'Naming Names: The Art of Memory and the NAMES Project AIDS Quilt.' *Critical Inquiry* 19 (summer 1993): 757.
38 Hawkins, 'Naming Names', 760.
39 Steven J. Gambardella, 'Absent Bodies: The AIDS Memorial Quilt as Social Melancholia.' *Journal of American Studies* 45, no. 2 (2011): 214, 219.
40 Cvetkovich, *An Archive of Feelings*, pp. 5–9.
41 Cvetkovich, *An Archive of Feelings*, p. 10.
42 Susan Leigh Star and James R. Griesemer, 'Institutional Ecology, "Translations" and Boundary Objects: Amateurs and Professionals in Berkeley's Museum of Vertebrate Zoology, 1907–39.' *Social Studies of Science* 19 (1989): 389. Emphasis in original.
43 Leigh Star and Griesemer, 'Institutional Ecology', 393.
44 A claim that is central to my current dissertation project.
45 J. C. Flügel, *The Psychology of Clothes* (London: Hogarth Press and the Institute of Psycho-Analysis, 1971 [1930]), pp. 15–16.

46 Erving Goffman, *Stigma: Notes on the Management of Spoiled Identity* (Englewood Cliffs, NJ: Prentice-Hall, Inc., 1963), pp. 2–3.
47 Samantha King, *Pink Ribbons, Inc.: Breast Cancer and the Politics of Philanthropy* (Minneapolis, MN: University of Minnesota Press, 2006), p. x.
48 King, *Pink Ribbons, Inc.*, p. xxx.
49 King, *Pink Ribbons, Inc.*, p. xxii.
50 King, *Pink Ribbons, Inc.*, p. xxiii. Emphasis added.
51 Komen Puget Sound, 'Race for the Cure raises more than $500,000 for Susan G. Komen Puget Sound.' https://komenpugetsound.org/race-for-the-cure-raises-more-than-500000-for-susan-g-komen-puget-sound/.
52 Marc Richardson, 'A History of the T-Shirt in Protest, Politics and Activism.' *Grailed*, 11 June 2020. www.grailed.com/drycleanonly/political-tshirt-history.
53 Angela Taylor, 'The T-Shirt Has Become the Medium for a Message.' *New York Times*, 17 August 1973, p. 36.
54 The ad, perhaps tellingly, includes a shirt with a pink ribbon on it. Bonfire, 'Make a Difference with T-Shirt Fundraising.' YouTube video, 0:30, 16 April 2020. www.youtube.com/watch?v=aRoTGdDt_cs.
55 T-ShirtKings247, 'Dayton Custom Memorial T-Shirts and RIP Shirts.' https://tshirtkings247.com/collections/frontpage.
56 T-ShirtKings247, 'Dayton Custom Memorial T-Shirts and RIP Shirts.'
57 T-ShirtKings247, 'Dayton Custom Memorial T-Shirts and RIP Shirts.'
58 Beverly Greene, 'Psychotherapy with African American Women: Integrating Feminist and Psychodynamic Models.' *Smith College Studies in Social Work* 67, no. 3 (1997): 304.
59 Ronald C. Fudge, 'The Use of Behavior Therapy in the Development of Ethnic Consciousness: A Treatment Model.' *Cognitive and Behavioral Practice* 3 (1996): 325.
60 Fudge, 'The Use of Behavior Therapy in the Development of Ethnic Consciousness', 328.
61 Eliott C. McLaughlin, 'RIP Tees: In Murder City, a Market for Wearable Memorials.' CNN, 1 March 2012. https://news.blogs.cnn.com/2012/03/01/rip-tees-in-murder-city-a-market-for-wearable-memorials/.
62 Chevel Johnson, 'T-Shirts Honor Casualties of Mean Streets.' *Los Angeles Times*, 11 October 1998. www.latimes.com/archives/la-xpm-1998-oct-11-mn-31353-story.html.
63 Gallagher, 'These T-Shirts Are Raising Serious Funds.'
64 King, *Pink Ribbons, Inc.*, pp. xi, 39.
65 Katie Mettler, 'Gucci's Straitjackets Draw a Model's Silent Protest on the Runway: "Mental Health Is Not Fashion."' *Washington Post*, 23

September 2019. www.washingtonpost.com/lifestyle/2019/09/23/guc
cis-straitjackets-draw-models-silent-protest-runway-mental-health-is-
not-fashion/.

66 Vanessa Friedman, 'Before Sex, the Straitjacket?' *New York Times*,
23 September 2019. www.nytimes.com/2019/09/23/style/gucci-ale
ssandro-michele-milan-fashion-week.html.

67 Tan-Jones uses they/them pronouns. Hannah Marriott, 'Gucci Model
Stages Mental Health Protest at Milan Fashion Week.' *The Guardian*,
22 September 2019. www.theguardian.com/fashion/2019/sep/22/
gucci-model-mental-health-protest-milan-fashion-week.

68 Novelist and scholar Alison Lurie, for instance, has argued for a
semiotics of dress. Lurie wrote that 'For thousands of years human
beings have communicated with one another first in the language of
dress', one with an associated vocabulary and grammar. 'To choose
clothes', Lurie contended, 'is to define and describe ourselves'.
Alison Lurie, *The Language of Clothes* (New York: Random House,
1981), pp. 3–5. Lurie cites novelist Honoré de Balzac and literary
theorist Roland Barthes as predecessors in this area, and others have
expanded on Lurie's formulations. See Malcolm Barnard, *Fashion as
Communication* (London: Routledge, 1996).

69 Taylor, 'The T-Shirt Has Become the Medium for a Message', p. 36.

70 Nancy Tomes, *Remaking the American Patient: How Madison Avenue
and Modern Medicine Turned Patients into Consumers* (Chapel Hill,
NC: University of North Carolina Press, 2016), p. 8.

71 *The American Heritage Dictionary of the English Language*, 5th
edition. 'Consumerism' (Boston, MA: Houghton Mifflin Harcourt,
2020).   www.ahdictionary.com/word/search.html?q=consumerism.
Emphasis in original.

72 D.W. Winnicott, 'Transitional Objects and Transitional Phenomena: A
Study of the First Not-Me Possession.' *International Journal of Psycho-
Analysis* 34, no. 2 (1953): 89–97.

73 On the first point, see Belk, 'Possessions and the Extended Self.'

74 Bonnie Honig, *Public Things: Democracy in Disrepair* (New York:
Fordham University Press, 2017), pp. 16–17.

75 Sanders, 'Memorial T-Shirts Create a Little Justice.'

76 Michael Pina, 'Is It OK to Sell Breonna Taylor T-Shirts?' *GQ*, 28
August 2020. www.gq.com/story/breonna-taylor-tshirt-ethics.

77 Pina, 'Is It OK to Sell Breonna Taylor T-Shirts?'

78 Quincy Snowdon, 'Elijah McClain Tragedy, Illustration and Anger
Ripples across Social Media.' *Sentinel*, 15 June 2020. https://senti
nelcolorado.com/news/metro/elijah-mcclain-tragedy-illustration-and-
anger-ripples-through-social-media/.

79 Sheneen McClain, update on www.gofundme.com/f/elijah-mcclain from 26 February 2021.
80 Justin Ray, 'Commentary: We Need to Talk about Those Breonna Taylor T-shirts.' *Los Angeles Times*, 28 September 2020. www.lati mes.com/entertainment-arts/story/2020–09–28/black-lives-matter-problem-with-breonna-taylor-t-shirts.
81 Josh Peter, 'Cashing in on George Floyd: T-Shirts, Pillows, Running Shoes and Even Underwear Are Being Sold, Some of It through Amazon.' *USA Today*, 15 June 2020. www.usatoday.com/story/money/2020/06/15/george-floyd-death-protests-lead-merchandise-sales-amazon/5337489002/.
82 Jim Dalrymple II, 'Michael Brown's Family Got into a Huge Brawl over Memorial T-Shirt Sales.' Buzzfeed News, 6 November 2014. www.buzzfeednews.com/article/jimdalrympleii/michael-browns-fam ily-got-into-a-huge-brawl-over-memorial-t.
83 Thank you to members of Harvard's History of Medicine Working Group for pointing me in this direction. See Texas Heart Institute, 'Wallace D. Wilson Museum.' www.texasheart.org/the-institute/museum/.
84 Ray, 'Commentary.'
85 Brittany Kriegstein, 'Memorial T-Shirts Bring Business to Bronx Shop, Comfort Grieving Families.' *New York Daily News*, 30 August 2020. www.nydailynews.com/new-york/ny-memorial-t-shirts-bring-busin ess-to-this-bronx-shop-20200830–4itexxw7vbeuxkaqwm5xobz3ta-story.html.
86 Ali Rowhani-Rahbar, Duane Alexander Quistberg, Erin R. Morgan, Anjum Hajat, and Frederick P. Rivara, 'Income Inequality and Firearm Homicide in the United States.' *Injury Prevention* (blog), *BMJ Injury Prevention*, 5 September 2019. https://blogs.bmj.com/injury-prevent ion/2019/09/05/income-inequality-and-firearm-homicide-in-the-uni ted-states/. Emphasis in original.
87 Ray, 'Commentary.'
88 McClain, update on www.gofundme.com/f/elijah-mcclain.
89 Minuchin, 'Ecology and Child Therapy', 55.
90 See Center for Behavioral Health Statistics and Quality, Racial/Ethnic Differences in Mental Health Service Use among Adults and Adolescents (2015–2019), Publication No. PEP21–07–01–002 (Rockville, MD: Substance Abuse and Mental Health Services Administration, 2021). Retrieved from www.samhsa.gov/data/.
91 David J. Rothman, *Strangers at the Bedside: A History of How Law and Bioethics Transformed Medical Decision Making* (New York: Aldine de Gruyter, 2003 [1991]).

92 Grant Stringer, 'Family of Man Who Died after Aurora Police Encounter Say It Was "Cold-Blooded Murder".' *Sentinel*, 2 October 2019. https://sentinelcolorado.com/news/metro/family-of-man-who-died-after-aurora-police-encounter-say-it-was-cold-blooded-murder/.
93 See www.amazon.com/Mcclain-Classic-Trending-Graphic-Fashiona ble/dp/B08BTWJYXN?th=1&psc=1.
94 Siegel, 'Honoring Loved Ones with "Rest in Peace" Shirts.'

Bibliography

The American Heritage Dictionary of the English Language, 5th edition. 'Consumerism.' Boston, MA: Houghton Mifflin Harcourt, 2020. www.ahdictionary.com/word/search.html?q=consumerism

Aronowitz, Robert A. *Unnatural History: Breast Cancer and American Society*. New York: Cambridge University Press, 2007.

Bailey, Zinzi D., Nancy Krieger, Madina Agénor, Jasmine Graves, Natalia Lionos, and Mary Bassett. 'Structural Racism and Health Inequities in the USA: Evidence and Interventions.' *Lancet* 389 (2017): 1453–63.

Barnard, Malcolm. *Fashion as Communication*. London: Routledge, 1996.

Bayer, Ronald. *Homosexuality and American Psychiatry: The Politics of Diagnosis*. New York: Basic Books, 1981.

Belk, Russell W. 'Possessions and the Extended Self.' *Journal of Consumer Research* 15, no. 2 (1998): 139–68.

Benjamin, Ruha. 'Conjuring Difference, Concealing Inequality: A Brief Tour of Racecraft.' *Theory and Society* 43, no. 6 (2014): 683–8.

Bijker, Wiebe E., and John Law (eds). *Shaping Technology/Building Society: Studies in Sociotechnical Change*. Cambridge: MIT Press, 1992.

Bonfire. 'Make a Difference with T-Shirt Fundraising.' YouTube video, 2020, 0:30. www.youtube.com/watch?v=aRoTGdDt_cs.

Boyd-Franklin, Nancy. *Black Families in Therapy: Understanding the African American Experience*. New York: Guilford Press, 2003.

Brooks, Robin. 'R.I.P. Shirts *or* Shirts of the Movement: Reading the Death Paraphernalia of Black Lives.' *Biography* 41, no. 4 (2018): 807–30.

Center for Behavioral Health Statistics and Quality. *Racial/Ethnic Differences in Mental Health Service Use among Adults and Adolescents (2015–2019)*, Publication No. PEP21–07–01–002. Rockville, MD: Substance Abuse and Mental Health Services Administration, 2021. Retrieved from www.samhsa.gov/data/.

Cvetkovich, Ann. *An Archive of Feelings: Trauma, Sexuality, and Lesbian Public Cultures*. Durham, NC: Duke University Press, 2003.

Dalrymple II, Jim. 'Michael Brown's Family Got into a Huge Brawl over Memorial T-Shirt Sales.' Buzzfeed News, 2014. https://web.archive.org/web/20210304160326/www.buzzfeednews.com/article/jimdalrympleii/michael-browns-family-got-into-a-huge-brawl-over-memorial-t.

Epstein, Steven. *Impure Science: AIDS, Activism, and the Politics of Knowledge*. Berkeley, CA: University of California Press, 1996.

Fleetwood, Nicole R. *Troubling Vision: Performance, Visuality, and Blackness*. Chicago, IL: University of Chicago Press, 2011.

Flügel, J.C. *The Psychology of Clothes*. London: Hogarth Press and the Institute of Psycho-Analysis, 1930.

Friedman, Vanessa. 'Before Sex, the Straitjacket?' *New York Times* , 2019. https://web.archive.org/web/20200717130317/www.nytimes.com/2019/09/23/style/gucci-alessandro-michele-milan-fashion-week.html.

Fudge, Ronald C. 'The Use of Behavior Therapy in the Development of Ethnic Consciousness: A Treatment Model.' *Cognitive and Behavioral Practice* 3 (1996): 317–35.

Gallagher, Jacob. 'These T-Shirts Are Raising Serious Funds for Antiracism Causes.' *Wall Street Journal*, 2020. https://web.archive.org/web/20210621204309/www.wsj.com/articles/these-t-shirts-are-raising-serious-funds-for-black-aligned-causes-11592335526.

Gambardella, Steven James. 'Absent Bodies: The AIDS Memorial Quilt as Social Melancholia.' *Journal of American Studies* 45, no. 2 (2011): 213–26.

Goffman, Erving. *Stigma: Notes on the Management of Spoiled Identity*. Englewood Cliffs, NJ: Prentice-Hall, 1963.

Greene, Beverly. 'Psychotherapy with African American Women: Integrating Feminist and Psychodynamic Models.' *Smith College Studies in Social Work* 67, no. 3 (1997): 299–322.

Hawkins, Peter S. 'Naming Names: The Art of Memory and the NAMES Project AIDS Quilt.' *Critical Inquiry* 19 (1993): 752–79.

Honig, Bonnie. *Public Things: Democracy in Disrepair*. New York: Fordham University Press, 2017.

Hundreds, Bobby. *This Is Not a T-Shirt: A Brand, A Culture, A Community – A Life in Streetwear*. New York: Farrar, Straus and Giroux, 2019.

Johnson, Chevel. 'T-Shirts Honor Casualties of Mean Streets.' *Los Angeles Times*, 1998. www.latimes.com/archives/la-xpm-1998-oct-11-mn-31353-story.html.

King, Samantha. *Pink Ribbons, Inc.: Breast Cancer and the Politics of Philanthropy*. Minneapolis, MN: University of Minnesota Press, 2006.

Komen Puget Sound. 'Race for the Cure raises more than $500,000 for Susan G. Komen Puget Sound.' https://web.archive.org/web/20210228060604/https://komenpugetsound.org/race-for-the-cure-raises-more-than-500000-for-susan-g-komen-puget-sound/.

Kranzberg, Melvin. 'At the Start.' *Technology and Culture* 1, no. 1 (1959): 1–10.

Kriegstein, Brittany. 'Memorial T-Shirts Bring Business to Bronx Shop, Comfort Grieving Families.' *New York Daily News*, 2020. https://web.archive.org/web/20220919213234/www.nydailynews.com/new-york/ny-memorial-t-shirts-bring-business-to-this-bronx-shop-20200830–4itexxw7vbeuxkaqwm5xobz3ta-story.html.

Löwy, Ilana (2010). *Preventive Strikes: Women, Precancer, and Prophylactic Surgery*. Baltimore, MD: Johns Hopkins University Press, 2010.

Lurie, Alison. *The Language of Clothes*. New York: Random House, 1981.

Marriott, Hannah. 'Gucci Model Stages Mental Health Protest at Milan Fashion Week.' *The Guardian*, 2019. https://web.archive.org/web/20210120214140/www.theguardian.com/fashion/2019/sep/22/gucci-model-mental-health-protest-milan-fashion-week.

McLaughlin, Eliott C. 'RIP Tees: In Murder City, a Market for Wearable Memorials.' CNN, 2012. https://web.archive.org/web/20210324200345/https://news.blogs.cnn.com/2012/03/01/rip-tees-in-murder-city-a-market-for-wearable-memorials/.

Mettler, Katie. 'Gucci's Straitjackets Draw a Model's Silent Protest on the Runway: "Mental Health Is Not Fashion."' *Washington Post*, 2019. https://web.archive.org/web/20210228080236/www.washingtonpost.com/lifestyle/2019/09/23/guccis-straitjackets-draw-models-silent-protest-runway-mental-health-is-not-fashion/.

Metzl, Jonathan Michel. *The Protest Psychosis: How Schizophrenia Became a Black Disease*. Boston, MA: Beacon Press, 2010.

Metzl, Jonathan Michel. *Prozac on the Couch: Prescribing Gender in the Era of Wonder Drugs*. Durham, NC: Duke University Press, 2003.

Metzl, Jonathan Michel and Helena Hansen. 'Structural Competency: Theorizing a New Medical Engagement with Stigma and Inequality.' *Social Science & Medicine* 103 (2014): 126–33.

Minuchin, Patricia, Jorge Colapinto, and Salvador Minuchin. *Working with Families of the Poor*, 2nd edition. New York: Guilford Press, 1998.

Minuchin, Salvador. 'The Use of an Ecological Framework in the Treatment of a Child', in *The Child in His Family*, Vol. 1, edited by E. James Anthony and Cyrille Koupernik, pp. 41–57. New York: Wiley, 1970.

Minuchin, Salvador. *Families and Family Therapy*. Cambridge, MA: Harvard University Press, 1974.

Minuchin, Salvador. 'Deconstructing Minuchin.' *Journal of Systemic Therapies* 36, no. 4 (2017): 95–7.

National AIDS Memorial. 'The History of the Quilt.' https://web.archive.org/web/20210316135241/www.aidsmemorial.org/quilt-history.

O'Neill, Katie Kavanagh. 'Mobtown Memories: Towards a People's History of Violence in Baltimore.' PhD dissertation, University of Pittsburg, 2017.

Peter, Josh. 'Cashing in on George Floyd: T-Shirts, Pillows, Running Shoes and Even Underwear Are Being Sold, Some of It through Amazon.' *USA Today*, 2020. https://web.archive.org/web/20210310211751/www.usatoday.com/story/money/2020/06/15/george-floyd-death-protests-lead-merchandise-sales-amazon/5337489002/.

Pina, Michael. 'Is It OK to Sell Breonna Taylor T-Shirts?' *GQ*, 2020. https://web.archive.org/web/20210301164208/www.gq.com/story/breonna-taylor-tshirt-ethics.

Ray, Justin. 'Commentary: We Need to Talk about Those Breonna Taylor T-shirts.' *Los Angeles Times*, 2020. https://web.archive.org/web/202 10317010234/www.latimes.com/entertainment-arts/story/2020–09– 28/black-lives-matter-problem-with-breonna-taylor-t-shirts.

Richardson, Marc. 'A History of the T-Shirt in Protest, Politics and Activism.' *Grailed*, 2020. https://web.archive.org/web/20220116111 139/www.grailed.com/drycleanonly/political-tshirt-history.

Rothman, David J. *Strangers at the Bedside: A History of How Law and Bioethics Transformed Medical Decision Making.* New York: Aldine de Gruyter, 1991.

ross, kihana miraya. 'Call It What It Is: Anti-Blackness.' *New York Times*, 2020. https://web.archive.org/web/20210315175728/www.nyti mes.com/2020/06/04/opinion/george-floyd-anti-blackness.html.

Rowhani-Rahbar, Ali, Duane Alexander Quistberg, Erin R. Morgan, Anjum Hajat, and Frederick P. Rivara. 'Income Inequality and Firearm Homicide in the United States.' *Injury Prevention* (blog), *BMJ Injury Prevention*, 2019. https://web.archive.org/web/20210412223215/ https://blogs.bmj.com/injury-prevention/2019/09/05/income-inequality- and-firearm-homicide-in-the-united-states/.

Sanders, Jasmine. 'Memorial T-Shirts Create a Little Justice, a Tiny Peace.' *New York Times*, 2017. https://web.archive.org/web/20210205020119/ www.nytimes.com/2017/11/14/style/memorial-t-shirts.html.

Siegel, Rachel. 'Where Gun Violence Abounds, Honoring Loved Ones with "Rest in Peace" Shirts.' *Washington Post*, 2019. https://web.archive. org/web/20200312223546/www.washingtonpost.com/business/2019/ 08/10/mass-shootings-everyday-grief-honoring-loved-ones-with-rest- peace-shirts/.

Silva, Daniella, and Ed Ou. 'Three Months after George Floyd's Killing, Memorial Remains "Sacred place" for Racial Justice.' NBC News, 2020. https://web.archive.org/web/20210320110711/www.nbcnews. com/news/us-news/three-months-after-george-floyd-s-killing-memorial- remains-sacred-n1238084.

Snowdon, Quincy. 'Elijah McClain Tragedy, Illustration and Anger Ripples across Social Media.' *Sentinel*, 2020. https://web.archive.org/web/202 10226214428/https://sentinelcolorado.com/news/metro/elijah-mcclain- tragedy-illustration-and-anger-ripples-through-social-media/.

Star, Susan Leigh, and James R. Griesemer. 'Institutional Ecology, "Translations" and Boundary Objects: Amateurs and Professionals in Berkeley's Museum of Vertebrate Zoology, 1907–39.' *Social Studies of Science* 19, no. 3 (1989): 387–420.

Stringer, Grant. 'Family of Man Who Died after Aurora Police Encounter Say It Was "Cold-Blooded Murder".' *Sentinel*, 2019. https://web.arch ive.org/web/20210926125743/https://sentinelcolorado.com/news/ metro/family-of-man-who-died-after-aurora-police-encounter-say-it- was-cold-blooded-murder/.

Summers, Martin. *Madness in the City of Magnificent Intentions: A History of Race and Mental Illness in the Nation's Capital*. New York: Oxford University Press, 2019.

Taylor, Angela. 'The T-Shirt Has Become the Medium for a Message.' *New York Times*, 1973.

Texas Heart Institute. 'Wallace D. Wilson Museum.' https://web.archive. org/web/20210421084444/www.texasheart.org/the-institute/museum/.

Tomes, Nancy. *Remaking the American Patient: How Madison Avenue and Modern Medicine Turned Patients into Consumers*. Chapel Hill, NC: University of North Carolina Press, 2016.

T-ShirtKings247. 'Dayton Custom Memorial T-shirts and RIP Shirts.' https://tshirtkings247.com/collections/frontpage. Accessed 24 February 2021.

Wailoo, Keith. *How Cancer Crossed the Color Line*. New York: Oxford University Press, 2012.

Weinstein, Deborah. *The Pathological Family: Postwar America and the Rise of Family Therapy*. Ithaca, NY: Cornell University Press, 2013.

Winnicott, D.W. 'Transitional Objects and Transitional Phenomena: A Study of the First Not-Me Possession.' *International Journal of Psycho-Analysis* 34, no. 2 (1953): 89–97.

Index

Hill, Austin Bradford 144, 158
HIV 63, 123, 128, 218
Honig, Bonnie 229
Howell, Joel 7, 43
Howell, Mary 200
Hunt, Nina 123
'hygienic sublime' 118

ideal patients 23
identified patient 220
 IP 220
Imperial Chemical Industries 194,
 204, 208, 215
informants 10, 113, 114, 116,
 119, 121, 123, 124, 125,
 126, 128
Institutional Review Board 126
 IRB 126
International Agency for Research
 on Cancer 202
International Coalition for Drug
 Awareness 138
*International Journal of Risk and
 Safety in Medicine* 152,
 159, 160, 161, 162, 163
internet 9, 10–11, 137, 140, 152,
 157, 161, 179
 web 2.0 148
 the web 138–41, 156
 World Wide Web 138

Jewson, Nicholas 3
Jofre, Shelley 146, 159, 163
Johnson, Charles 25
Jones, Cleve 223
*Journal of the American Dental
 Association* 21, 25,
 43, 49, 50
*Journal of the American Medical
 Association* 64
 JAMA 64, 68, 74, 77

Kaunitz, Karen 123
Kelley, William E. 88
Kennedy, Edward 57
kidney disease 53, 59, 71
Kidney Donor Risk Index 68
 KDRI 68

kidney failure 9, 53, 57, 58–9, 62,
 64, 67–8, 69–70
King, Samantha 224
Klawiter, Maren 197, 208, 209
Kolff, Willem 53
Kornblau, Barbara 87–9, 99
Kranzberg, Melvin 217, 233

Lancet 74, 78, 201
laparoscopic cholecystectomy
 81–2, 83–96, 101, 106
laparoscopy 81, 86, 91, 95, 97
laser surgery 86, 99, 101, 104, 106
Law, John 217, 233, 239
Leber, Paul 142
Linker, Beth 58, 73, 98
lithotripsy 85
Lochlann Jain, Sarah 115, 130
logocentrism 139
L'Oréal 175–6
 Ambre Solaire 176
 Decléor 176, 181
Los Angeles Times 94, 99, 102,
 105, 106, 230, 240, 242
lumpectomy 195, 197

malignant melanoma 167
 melanoma 177–9, 182, 192
mammographic screening 197, 206
 mammography 198
Mangin, Dee 140, 144, 146
Marie Claire 177
Marietta, Georgia 86, 98
Marshall, Sylvia 59
Martin, Trayvon 219
mastectomy 197
Matthews, Gene 123, 132
Mbembe, Achille 55, 72
McClain, Elijah 230, 232, 242
McClain, Sheneen 230
McKernan, Barry 86, 88, 98
Medawar, Charles 146, 159
Medicaid 58, 64, 73, 77
Medical Ambulatory Care 61
medical marketplace 3, 4
medicalization 4
Medicare 52, 57–8, 62, 64,
 73, 77, 79